THE DEVELOPMENT OF MICROBIOLOGY

THE DEVELOPMENT OF
MICROBIOLOGY

PATRICK COLLARD
MD, MSc, FRCP

PROFESSOR OF BACTERIOLOGY, UNIVERSITY OF MANCHESTER

CAMBRIDGE UNIVERSITY PRESS
CAMBRIDGE
LONDON · NEW YORK · MELBOURNE

Published by the Syndics of the Cambridge University Press
The Pitt Building, Trumpington Street, Cambridge CB2 1RP
Bentley House, 200 Euston Road, London NW1 2DB
32 East 57th Street, New York, NY 10022, USA
296 Beaconsfield Parade, Middle Park, Melbourne 3206, Australia

First published 1976

Printed in Great Britain
at the
University Printing House, Cambridge
(Euan Phillips, University Printer)

Library of Congress Cataloguing in Publication Data

Collard, Patrick.

The development of microbiology.

Includes bibliographies and index.

1. Microbiology – History. I. Title.
QR21.C64   576'.09   75-40987
ISBN 0 521 21177 8

CONTENTS

*For the Manchester Diploma in Bacteriology students
who persuaded me to write this book*

PREFACE

Science is a continuously developing system: a series of models each replacing a previous less satisfactory one. If we are to understand the contemporary corpus of knowledge and critically assess future developments we must learn how we have arrived at the present position.

This book based on lectures given to postgraduate students is an attempt to tell the story of the development of certain ideas in microbiology, relating the views held at different times to the contemporaneous state of knowledge in other fields and showing how successive models grew out of the internal contradictions of their predecessors.

My thanks are due to Mr D. F. Cook and the staff of the Manchester University Medical Library for unstinting help in obtaining references and to my secretary Miss Brenda Gardner for her unfailing patience during the preparation of the typescript for the press.

November 1975   Patrick Collard

1

Introduction

Microbiology, like all the sciences, is founded upon the twin pillars of craft techniques and philosophical speculation. Without the empirical observations of the first, the subject would be but a mass of meaningless verbiage, and without the organizing hypotheses of the second, would be but a collection of descriptions and receipts.

The crafts of food preservation and fermentation have been highly developed for many thousands of years. Early Egyptian papyri contain detailed instructions for the brewing of wine and beer, and it is clear from these documents that the importance of excluding air from the secondary fermentation was well recognized. The principle of using a starter was standard practice in the making of leavened bread and as the use of the deposit from fermented beer as an agent for the raising of dough is mentioned, it is clear that the Ancient Egyptians were aware of the identity of the agent that produced these two processes, although of course they were quite unaware of its nature.

Other microbiological processes such as the retting of flax are also of great antiquity and had been brought to a high degree of technical excellence three or four thousand years ago.

The use of microbes to produce various milk products such as cheese and the various sour milk drinks, yoghurt and kamous, probably goes back to the time of the neolithic agricultural revolution when men first domesticated grazing animals and began to tend them in herds.

The preservation of foodstuffs by methods such as drying, salting, and dehydration by immersing them in strong sugar solutions, seem to have come in early in the neolithic period, as soon as men had a surplus of foodstuffs sufficient to last them from harvest to harvest.

While the craft techniques have been known for several thousand years, the speculation of the philosophers at first lagged behind. The phenomena of fermentation, putrefaction and infectious diseases were well recognized, but the explanations advanced to account for them were all unsatisfactory because they were grounded in a world view that was pre-scientific and often magical, and within such a framework the true explanation of these phenomena as by-products of microbial growth was inconceivable.

Formal microbiology may be considered as having passed through four eras. Firstly the era of speculation, lasting from about 5000 BC to 1675. Secondly the era of observation, lasting from 1675 to the mid nineteenth century. Thirdly the era of cultivation, lasting from the mid nineteenth century until the beginning of the twentieth century. Fourthly the era of physiological study, commencing at the turn of the century and continuing to the present.

During the long era of speculation several thinkers advanced the hypothesis that contagious diseases might be due to the growth of minute living organisms, but in the absence of microscopes their theories could not be put to the test and were therefore, in the words of the logical positivist school of philosophers, 'meaningless statements'. In the classical period Cicero discussed the possibility of fevers being caused by the multiplication of minute animals, and some fifteen hundred years later the Renaissance scholar Fracastorius wrote of a '*contagium vivum*', but neither of these hypotheses was fruitful in the circumstances prevailing when they were advanced.

The era of observation dates from the work of the Dutch microscopist Antony van Leeuwenhoek, first published in 1675. He was undoubtedly the first man to see and describe bacteria. His drawings are still extant and so clear are they that one can recognize bacilli, streptococci and other characteristic forms. The era of observation continued for nearly two hundred years and although numerous microscopic forms of life were described, some bacteria, some protozoa and some fungi, and known to their discoverers as 'animalcules', knowledge of the function of these micro-organisms was not advanced at all. They remained scientific curiosities and nothing more. Such progress as was made during this period was the result of a continuing controversy concerning the possibility or otherwise of spontaneous generation. The scholastic philosophers of the Middle Ages were convinced of the reality of spontaneous generation and regarded it as a common phenomenon. As late as the early seventeenth century van Helmont gave a recipe for the spontaneous generation of mice from grain, but during the years of the enlightenment men ceased to believe in the spontaneous generation of whole animals, as a result of the experimental work of men such as Francesco Redi (1626–1697) who showed unequivocally that maggots were not produced by spontaneous generation from rotting meat, as was the general belief at the time, but appeared if, and only if, adult flies laid their eggs in the meat. Redi's experiments, which involved protecting fresh meat from contamination by flies by means of a gauze shield, were elegant, simple and conclusive. William Harvey, the discoverer of the circulation of the blood, summed up the opinion of the 'anti-spontaneous-generationists' in an aphorism '*Omnia ex ovo*'. There were, however, a number of scholars who maintained that spontaneous generation took place, at least in the case of the animalcules of van Leeuwenhoek, and in the effort to decide whether or not this was the case much was learned about the behaviour of microbes, and many of the techniques which were later to be perfected as the basis of cultural bacteriology were first developed.

The most famous of these controversies was that carried on in the mid eighteenth century between Father Joseph Needham and the Italian priest Lazaro Spallanzani, a professor at the University of Pavia. Needham claimed to have shown that microbes were generated in samples of broth that had been boiled and then sealed in the vessels that had been used for the boiling. His work convinced the great French naturalist Buffon, but was challenged by Spallanzani who conducted further and better experiments in which he continued the boiling for much longer periods and carried out the boiling in previously sealed vessels, thus initiating the use of steam under pressure as a sterilizing agent. His work showed that it was possible to preserve broth indefinitely, so long as it had been heated to a temperature above the boiling point of water and was preserved from contamination by the outside air. An interesting by-product of these experiments to refute the doctrine of spontaneous generation was the first observation of anaerobic micro-organisms. The significance of Spallanzani's observations was not appreciated at the time and anaerobiosis had to be redis-covered by Louis Pasteur a hundred years later, but it is quite clear that the first observations ever made were his.

The era of cultivation was initiated by Louis Pasteur's studies on the nature of fermentations. Commencing in 1857 with the 'Mémoire su la fermentation apelée lactique' the work was extended to cover studies on butyric fermentations, fermenta-tions producing ethanol and the production of vinegar. Pas-teur's hypothesis that each type of fermentation was caused by the growth and metabolism of a specific micro-organism forced him to develop methods for the cultivation of each organism uncontaminated by any other species. In the early years he used very tedious liquid dilution methods to obtain pure cultures, but later used the methods of isolation on solid media developed by Robert Koch. In order to carry out this work Pasteur had to develop methods for the sterilization of his media and glassware as well as aseptic techniques for carrying out the dilutions and subcultures. The use of the naked flame, the hot air oven and the autoclave all originated in Pasteur's laboratory. Pasteur's work on fermentation started as an attempt to solve a specific industrial problem and this concern with the application of microbiology led him to publish his works on the diseases of wine and the diseases of beer. It is important to remember that the concept of the specific action of microbes was first developed as a result of what we should today call 'trouble shooting' in the fermentation industry. Pasteur was next asked to turn his attention to a disease which threatened to ruin the French silk industry. Silkworms were dying in large numbers and no method for the control of the outbreaks had been evolved. Pasteur's studies on this condition led him to the conclusion that the disease (pébrine) was caused by a protozoan parasite which he was able to demonstrate by microscopy in diseased silkworms but not in healthy ones. A policy of segregation of healthy worms and destruction of colonies which showed signs of infection brought the disaster under control. Pasteur's successes in controlling the diseases of wine and beer and the silkworm disease led to his being asked by the French Government to investigate the problem of anthrax, a condition that was causing great losses to livestock in the country. The work on anthrax and fowl cholera led to the concept of attenuation of bacterial cultures and the development of effective live vaccines for the control of both these infections. The culmination of Pasteur's work was the extension of these principles to the case of rabies, where he was able to produce an effective attenuated vaccine without at any time isolating the virus, and indeed without any idea of the nature of the causative organism.

The great contributions of Robert Koch to cultural bacteriology were the development of solid media and the technique of sorting out mixed cultures and obtaining pure single species cultures by streaking out onto the solid media. This technique revolutionized cultural bacteriology and made possible the great flowering of the subject during the last two decades of the nineteenth century. Between 1882 and 1900 the causal organisms of almost all of the bacterial diseases were isolated and effective preventive measures became possible.

At the same time as the foundations of hygienic bacteriology were being laid by the isolation of the causal organisms of the major epidemic diseases prevalent in Europe, the basis of chemotherapy and immunology was being developed by workers such as Paul Ehrlich, Elie Metchnikoff and Pfeiffer. Ehrlich's earliest work was concerned with the differential staining of leucocytes. The specific way in which certain cells took up different dyes suggested to him the possibility of exploiting this property for the differential destruction of parasites in the body of the host. His preliminary work on the treatment of malaria with methylene blue led ultimately to the synthesis of effective chemotherapeutic compounds such as trypan red and salvarsan.

The search for the basis of immunity followed two lines; the examination of the sera of immune and non-immune animals and the study of the cellular responses to infection. For many years there was a continuing controversy between the followers of Metchnikoff who adhered to the cellular hypothesis and the German scholars who believed in a humoral basis for immunity. Finally in 1908 the work of Wright and Douglas reconciled the two hypotheses and showed them both to be correct.

During the last decade of the nineteenth century great advances were made in our knowledge of the bacteria in soil and water which are responsible for the completion of the nitrogen, sulphur and carbon cycles, and thus for the continued fertility of the land. As a result of the work of Winogradski and Beijerinck a whole new world of bacteria became known, organisms with previously unheard of types of nutrition able to live and grow when supplied only with elementary nitrogen, iron or sulphur and carbon dioxide; the autotrophic bacteria.

The study of bacterial physiology and biochemistry that was the dominant theme of the subject during the twentieth century was stimulated by the isolation of cell-free enzymes and the consequent development of dynamic biochemistry which took place both in Germany and England at the beginning of the century. Perhaps the most famous names in the field of microbial biochemistry are those of Marjorie Stephenson who elucidated many of the energy-yielding pathways in her laboratory in Cambridge and Kluyver of Delft who had the genius to see the underlying unity in the diversity of microbial energy-yielding mechanisms. The detailed study of synthetic pathways came twenty years later and arose out of the application of biochemical genetics, originally worked out using fungi, by Beadle and his school in California.

The study of nutritionally deficient mutants not only disclosed the stepwise pathways by which complex molecules are synthesized in bacteria, but also led to a deeper understanding of the mechanisms of gene expression summed up in the aphorism 'one gene one enzyme'. This stimulated further research in bacterial genetics, at first applied to the problem arising from the emergence of antibiotic-resistant strains of bacteria in clinical practice, but soon extended to more fundamental studies which demonstrated the existence not only of sexuality in bacteria, but of other more bizarre types of genetic exchange, some of which had been reported previously as cases of inexplicable variation but were now for the first time understood. From these studies in bacterial genetics there evolved the new subject of molecular genetics, students of which have solved the genetic code and given us deeper insight into biology than anyone before them.

Pari passu with these advances in fundamental biology a revolution in the treatment of infectious diseases was taking place. In 1935 Domagk working along the lines that had been laid down by Ehrlich, reported the synthesis of prontosil red, a dye which while almost non-toxic to mammals, killed streptococci, neisseria and various other bacteria at very low concentrations. Within the year, French workers had hydrolysed the molecule and shown that the antibacterial activity resided not in the chromophore group but in the colourless sulphanilamide portion of the molecule. As a result of these discoveries many hundreds of sulphonamide drugs have been developed which are extremely effective in the treatment of a number of infectious diseases.

The observation by Fleming in 1928 that the mould *Penicillium notatum* produced a *diffusable* antibacterial substance did not immediately lead to any therapeutic advance, for although it was shown to be a non-toxic substance for animals, it proved impossible to prepare extracts of sufficient purity in sufficient quantity for therapeutic tests to be carried out. The problem of extracting and purifying the antibacterial substance, now known as penicillin, was re-opened in 1939 by a team of brilliant chemists headed by Sir Howard Florey, the Professor of Pathology at Oxford. By then chemical techniques for separating organic molecules were so far advanced that the problem was solved in principle within a year and adequate amounts of sufficiently pure penicillin were made available for first animal, and, when these proved dramatically successful, for human therapeutic trials. Penicillin proved to be virtually non-toxic to man and to be active against Gram-positive organisms and spirochaetes in very low concentrations.

The industrial production of penicillin was followed by a worldwide hunt for other organisms producing antibiotics, which resulted in the discovery first of streptomycin and later of the many broad-spectrum antibiotics. In recent years a new vista has been opened up by the development of semi-synthetic antibiotics. These are made by adding certain substrates to the fermentation which force the antiobiotic-producing organism to modify the nucleus of the compound and thus allow synthetic side groups to be added later.

The study of virology is usually assumed to have commenced with the recognition by Iwanowski in 1892 that tobacco mosaic disease was caused by a filter-passing ultra-microscopic organism, and the demonstration six years later by Loeffler and Frosch that foot-and-mouth disease was caused by a similar organism. Pasteur's work on rabies antedates this, but at that time there was no concept of a class of ultra-microscopic filter-passing infective agents. Just as bacteriology was liberated by Robert Koch's invention of solid media so virology took great steps forward after Goodpasture introduced the fertile hen's egg as a culture medium in the mid nineteen-thirties, and reached its present state of maturity only after Enders introduced the use of tissue cultures with added antibiotics in 1949. Today, virologists are able to grow most viruses, to detect specific antibodies in the serum of patients and to produce effective vaccines against a number of virus diseases.

Microbiology has provided the knowledge that has enabled the industrialized countries of the world to bring all the major infective diseases under control, but it must not be forgotten that they are controlled, not eradicated, and it is only by the continued vigilance of public health bacteriologists that this control can be maintained. Industrial microbiology is a fast-developing field where fresh developments may be expected year by year. Fundamental microbiology is more and more part of molecular biology and in this field we may expect some of the most exciting developments in the next decades.

FURTHER READING

Sigerist, H. E. *A History of Medicine*. London: Oxford University Press (1951).

Bulloch, W. *The History of Bacteriology*. London: Oxford University Press (1938).

Ford, W. W. *Bacteriology* (Clio Medica Series). New York: Hafner Publishing Co. (1964).

2

The development of the microscope, staining methods and morphological description

The simple microscope, made up of a single lens of very short focal length, evolved from the magnifying lenses which have been in use since the days of antiquity. The earliest examples known are biconvex lenses made from gem stones and found in the Assyrian excavations of Lugard. Classical authors mention the use of spherical glass vessels filled with water as magnifiers. It was not, however, until the late seventeenth century that lenses of a sufficient quality for the observation of bacteria were produced.

The first man to see bacteria was the Dutch microscopist Antony van Leeuwenhoek (Fig. 2.1). Van Leeuwenhoek was a draper in the quiet town of Delft. He ground his own lenses and during his lifetime he made several dozen simple microscopes with short-focal-length lenses ground so accurately that they gave magnifications of about ×300. Similar apparatus had been described by Descartes earlier in the century, but the quality of the lenses in the earlier models did not permit magnifications great enough to visualize bacteria.

As can be seen in Fig. 2.2, the lens was fixed and the specimen was moved into focus by the use of a series of fine screws. The type of illumination used is not known with certainty; it may have involved the use of a convex mirror, as did the model described by Descartes, or it may have made use of a larger biconvex lens to focus the light onto the specimen. Almost certainly the light source must have been sunlight as no contemporary artificial light would have been able to achieve the intensity of illumination required when working at such high magnification.

van Leeuwenhoek examined rain water, well water, sea water, and water in which peppercorns had been infused. He reported

9

Fig. 2.1. Antony van Leeuwenhoek (1632–1723). From A. van Leeuwenhoek, *Arcana natura detecta* (1695). By courtesy of 'The Wellcome Trustees'.

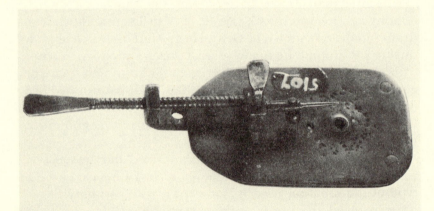

Fig. 2.2. The Leeuwenhoek microscope. A facsimile of the original in the University of Utrecht. By courtesy of 'The Wellcome Trustees'.

his observations to the Royal Society of London in a letter dated 9 October 1676. The letter is published in the *Philosophical Transactions of the Royal Society* for 1677, No. 133, pp. 821–31. A few years later in 1683 he contributed a further letter to the Society containing 'microscopical observations about animals in

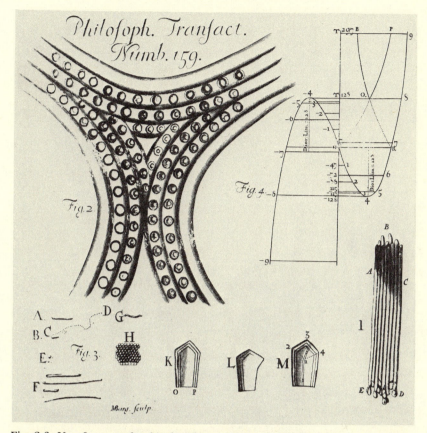

Fig. 2.3. Van Leeuwenhoek's drawings of bacteria. From *Phil. Trans. R. Soc.* (1684), **159**.

the scurf of the teeth and other matters', and an abstract of this accompanied by some of the author's drawings was published in the *Philosophical Transactions of the Royal Society* in 1684. These are the first published drawings of bacteria.

As can be seen in Fig. 2.3, the drawings are sufficiently accurate for us to be able to recognize streptococci, bacilli and spirochaetes. Not all of the organisms in the plates are bacteria, many of them are protozoa or fungi. It is of interest that van Leeuwenhoek was the first person to observe spermatozoa under the microscope.

It may at first appear odd that a Dutch draper living in Delft should publish his results in the *Philosophical Transactions of the Royal Society*. The explanation lies in the fact that at that period

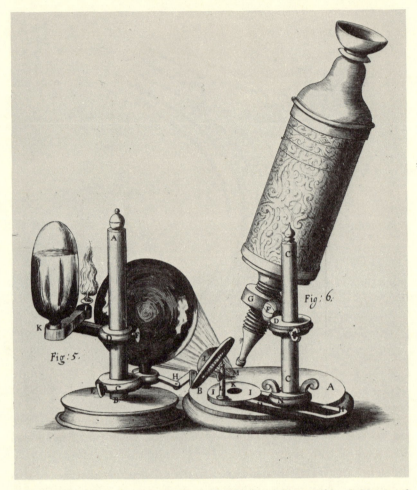

Fig. 2.4. Hooke's microscope. From R. Hooke, *Micrographia*. J. Martyn & J. Allestry, London (1665). By courtesy of 'The Wellcome Trustees'.

there were very close links between Holland and this country. The English king, Charles II, and members of his court had spent many years of exile in Holland, and it was they who had played a major part in the foundation of the Royal Society. There was at that time no similar society in Holland and so it was not surprising that van Leeuwenhoek should communicate his observations to friends who suggested publication in an English journal. The letter was originally in Dutch and was translated by the publisher. The publication of these observations naturally

Fig. 2.5. Giovanni Battista Amici
(1784–1863). From an original photograph
in the Wellcome Historical Medical Museum.
By courtesy of 'The Wellcome Trustees'.

generated great interest amongst the scientists of the time and led to van Leeuwenhoek's election as a Fellow of the Royal Society in 1680. It is noteworthy, however, that these observations did not lead to any investigation of the possible activities of the micro-organisms, either in fermentations or as possible causes of infectious disease. The state of development of chemistry and medicine was too primitive for such speculations to suggest themselves, and even had such questions occurred to scientists, they would have been unable to make any progress in answering them until methods of obtaining pure cultures had been developed.

Compound microscopes had first been made at the end of the sixteenth century in Germany, but the magnification of these instruments was too low for bacteria to be seen with them. During the seventeenth century compound microscopes with magnifications of ×300–×500 were constructed by Hooke and

Fig. 2.6. The working instrument of Professor Amici. A modification of his reflecting microscope for use with an achromatic objective. By courtesy of 'The Wellcome Trustees'.

others (Fig. 2.4), but with these higher powers came the problem of chromatic aberration. It proved impossible to make observations upon objects as small as bacteria as they were masked by the rings of light around them. During the eighteenth century many improvements in the mechanical construction of microscopes were made – fine controls for focusing, mechanical stages and more rigid tubes which improved the alignment of the lenses – and many of the later models such as the one made for King George III are collector's pieces of great elegance.

However, none of these instruments was capable of giving the resolution needed to obtain a clear picture of bacteria. Indeed, they were all inferior to the simple microscopes of van Leeuwenhoek from that point of view.

It was not until early in the nineteenth century that any progress was made. Professor Amici (Fig. 2.5) of the University of Medina, an eminent Italian mathematician, astronomer, physicist, and natural historian, was the first person to construct achromatic objective lenses. These were made by cementing together a biconvex and a biconcave lens of different refractive indices, thus markedly reducing the chromatic aberration and making possible good resolution at magnifications of up to 600 diameters. With these instruments (Fig. 2.6) it was just possible to see bacteria, but as no stains were in use it remained difficult to see more than an outline. Another limitation at this time and for many decades to come was the use of wet preparations. The organisms were not fixed and even non-motile cells were constantly undergoing Brownian movement, thus rendering accurate observation difficult. The first effective achromatic microscope was constructed by Amici in about 1821, and further improvements were made by the Frenchman, Chevalier and the Englishman, Lister. While the achromatic objectives of the early nineteenth century were a great improvement on anything that had been available before, they were by no means as good as the modern apochromatic objectives which were to be developed by the end of the century.

The staining of histological sections was first carried out by the German botanist, Ferdinand Cohn (Fig. 2.7) in 1849. At this stage only vegetable dyes were available and carmine and haematoxylin were used.

The first of the aniline dyes was synthesized from coal-tar in 1856 by Perkin of Manchester. The synthetic dyestuff industry was not, however, developed in England but in Germany where the level of technological education was much higher, and there were both manufacturers who saw the immense potentialities of the new methods and industrial chemists who were capable of developing the techniques required for large-scale production.

Fig. 2.7. Ferdinand Cohn (1828–1898). *From
Bull. Hist. Med.* (1929), **7**, facing p. 49.

The first report of the application of vegetable stains to
bacteria is that of Hoffman in 1869. He gives an account of the
use of carmine. In 1875 Weigert was using various simple stains,
amongst them the synthetic dye methylene blue, to stain
bacteria. It is important to remember that these early workers
stained the bacteria in suspension and examined wet preparations under the microscope.

In 1876 Robert Koch first described bacterial endospores.
The observations were made on *Bacillus anthracis* in hanging
drop preparations, and illustrated with line drawings (Fig. 2.8).

In the next year, 1877, Koch published an account of a
revolutionary advance in the microscopy of bacteria and illustrated it with the first photomicrographs of bacteria. Koch was
the first man to prepare dried films of bacteria and to stain them
with a dye, methylene blue. The films were dried in air and fixed
with alcohol, a technique developed by Paul Ehrlich for blood
films. The fixed and stained films were protected with a coverslip
and thus became permanent preparations (Fig. 2.9*a*).

The quality of the photomicrographs that Koch was able to

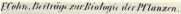

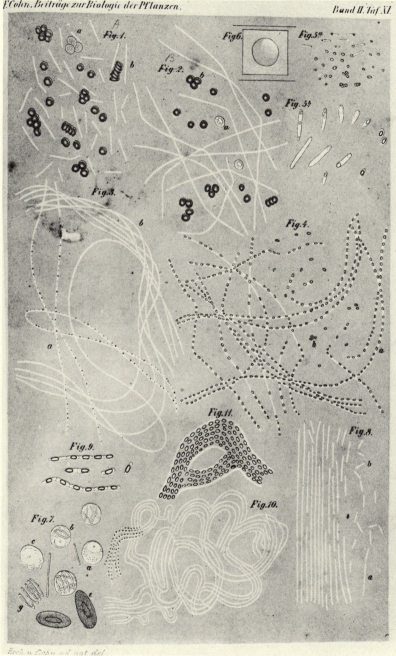

Fig. 2.8. Koch's line drawings of *Bacillus anthracis* in hanging drop preparations. From *Beitr. Biol. Pfl.* (1876–7), **2**.

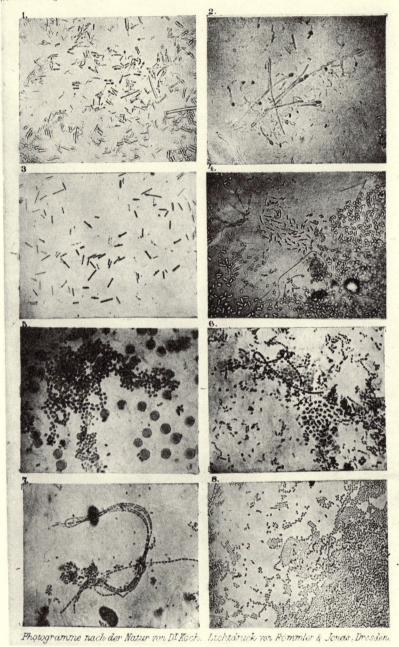

Photogramme nach der Natur von D.ͬ Koch. Lichtdruck von Römmler & Jonas, Dresden.

Fig. 2.9. Koch's photomicrographs of bacteria. Both from *Beitr. Biol. Pfl.*
(1876–7), **2**.

F.Cohn. Beiträge zur Biologie der Pflanzen. Band II. Taf. XVI.

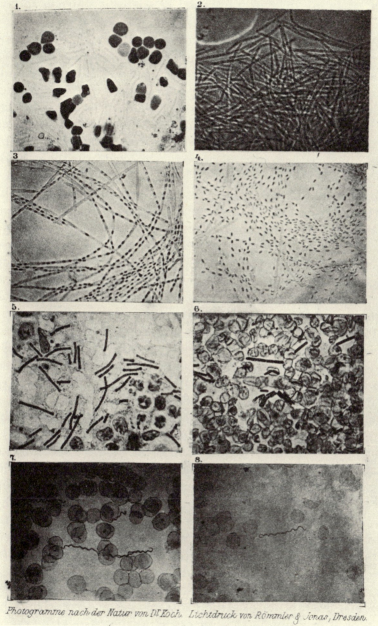

Photogramme nach der Natur von D.ᵣ Koch. Lichtdruck von Römmler & Jonas, Dresden.

produce is astonishing (Fig. 2.9*b*). These first pictures are as good as many that are published in scientific journals today. When we remember that photography in 1875 was still a matter of preparing one's own plates, using gelatine and a solution of silver salts, loading the plates into the camera in the dark and exposing the plate by taking off the cap over the lens by hand, timing the exposure with a stop-watch and finally developing and printing from the plates in a dark room that was an adapted kitchen or scullery, our admiration for this great pioneer becomes unbounded. There are very few other examples of techniques that have sprung already perfected from the mind of their originator.

The lenses that Koch used for these early photomicrographs were water immersion lenses. In 1878 Abbé of Zeiss introduced oil immersion lenses. This system, employing an oil of the same refractive index as the objective lens, made it possible to use objectives of the highest numerical aperture without being troubled by more than the slightest chromatic aberration and has been the standard method for critical observation at the highest magnifications possible with light microscopy ($\times 1000$–$\times 1200$) ever since. The greatest defect of the first oil immersion lenses was the apparent curvature of the field. While very good definition could be obtained at the centre of the field this was at the expense of having the periphery of the field out of focus.

Further improvements in the use of methylene blue for the staining of bacteria were published by Paul Ehrlich in 1881.

In 1882 Koch succeeded in staining the tubercle bacillus with alkaline methylene blue using heat to effect penetration of its waxy envelope. By this date the revolution in medical bacteriology was under way and all over Europe workers were attempting to demonstrate bacteria in sections of diseased tissues. The results with the simple stains then in use were unsatisfactory. With methylene blue the bacteria showed up as darker dots or rods against a blue background and it was often difficult to decide whether a dark blue dot was a bacterium or some artifact. The first attempt to solve the problem was by decolorizing the sections after staining. In some cases, notably tissues containing tubercle bacilli, this was effective, but in many cases the bacteria were as effectively decolorized as the tissues in which they were embedded and so nothing was gained. The problem was finally solved by Hans Christian Gram, the Danish pathologist who in 1884 introduced the technique of counter-staining after decolorization. In this method the bacteria stained a deep purple with gentian violet and the surrounding tissues with bismarck brown. Gram originally developed his staining technique for vizualizing bacteria in tissues, and for Gram-positive organisms it is a very effective method for doing so. Within a few years of the publication of the method it had become clear that it was of even greater value as a means of dividing bacteria into two classes according to their behaviour when the decolorizing agent was applied. Those that retained the violet dye were referred to as Gram-positive and those that were decolorized and so stained pink by the counter-stain were referred to as Gram-negative. From the first it was noticed that the Gram reaction was correlated with various other important physiological characters of bacterial species. It was not, however, until many years later that any understanding of the mechanism of the reaction was reached.

The staining of the tubercle bacilli had been achieved using methylene blue by Koch, and Paul Ehrlich in 1882 had employed methyl violet or alcoholic fuchsin in aniline water and nitric acid or hydrochloric acid as a decolorizing agent, but the results were not very satisfactory: too often the bacilli would be decolorized as well as the section and the results with films of cultures were even more uncertain. In the same year Ziehl introduced phenol, in place of aniline water, with methyl violet. In 1883 Neelsen described his modification of Ziehl's technique for the differential staining of tubercle bacilli, in which the bacterial cells were stained with hot carbol fuchsin and the decolorizing agent was 15 per cent sulphuric acid. The mycobacteria resist decolorization and stain red whilst tissue cells and other bacteria which are decolorized are counter-stained with methylene blue or malachite green. This method was found to be superior to anything then available and has become the standard method employed for staining tubercle bacilli. The method is usually referred to as that of Ziehl and Neelsen.

In 1890 flagella were first demonstrated by Frederick Loeffler (Fig. 2.10). The fact that some bacteria were motile and some were not had been known for many years, but the flagella of bacteria had not been seen under the microscope.

It was not until 1894 that the cell walls of bacteria were first

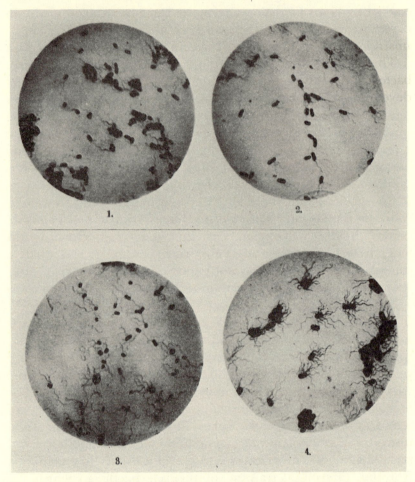

Fig. 2.10. The first photomicrograph of bacterial flagella, taken by Friedrich
Loeffler in 1890. From *Zentralbl.Bakt.* (1890), **7**.

demonstrated by Fischer. He used a technique of plasmolysis to
separate the cell membrane from the cell wall and thus was able
to show that these were two separate structures in bacteria.

By the end of the nineteenth century almost all the external
morphological features of bacterial cells were known, the
exception being the fimbriae which are not visible using the light
microscope. There was, however, no suspicion that bacteria
might possess nuclei: at this time bacteria were thought of as
simple cells with no internal organization. A few years later they
were described by a distinguished microbiologist as 'bags of

enzymes'. Nor was there any suggestion that such complex cell organelles as ribosomes were to be found in the apparently unstructured protoplasm of these simple cells.

The first step towards the exploration of the ultra-structure of bacteria and the visualization of virus particles came with the development of the ultra-violet microscope by Barnard in 1919. This instrument which used as ultra-violet light source, quartz lenses and a photographic system for recording the images, increased the resolution of the light microscope considerably and made possible the first photographs of the elementary bodies of a number of viruses. Up to this date only viral inclusion bodies had been seen, and they were of course intracellular colonies of virus particles embedded in a matrix of cell material. Although a number of elegant pictures of various bacterial cells were published using ultra-violet microscopy, no real advance in our knowledge of the structure of the bacterial cell resulted from its use.

In 1934 the Belgian physicist Marton built the first electron microscope. This type of instrument, using a beam of electrons as its source of 'illumination', a series of electromagnetic 'lenses' to focus the beam onto the specimen in its vacuum chamber, and photographic recording of the image so produced, increased the power of resolution of microscopes by several orders of magnitude. The limits of the conventional light microscope are reached at magnifications of about ×1200, and of the ultra-violet microscope at about ×2500, whilst the electron microscope can give good resolution at magnifications of ×200000–300000. The first electron microscope photographs of bacterial cells were published by Mudd and his colleagues in 1941. With the development of techniques for cutting ultra-thin sections, a whole new world was opened up to students of cellular morphology. The presence and peculiar nature of the bacterial nucleus became clear, intracellular organelles such as the ribosomes and the mesosomes were recognized, and the structural differences in the cell wall that determine the Gram staining of bacteria were elucidated. In addition, the presence of a new class of extracellular appendages, the fimbriae, was demonstrated and later this led to the recognition of the specialized fimbriae, known as pili, involved in the process of conjugation.

In 1957 E. C. Dougherty of Berkeley, California, published a short article in the *Journal of Protozoology* entitled 'Neologisms needed for the structures of primitive organisms': in it he proposed the terms procaryotic and eucaryotic to denote the types of cellular organization of the bacteria and blue-green algae on the one hand, and all other cells on the other. The importance of this evolutionary discontinuity cannot be over-emphasized; beside it even the differences between animals and green plants are overshadowed. It could only be recognized in the light of the electron microscope ultra-thin section studies that had been made over the previous seventeen years. Dougherty's paper *will* come to be accepted as one of the great synthetic contributions to biology. The exploration of subcellular structure of bacteria using higher magnifications continues; it seems likely that in a few years time it will be possible to obtain resolutions great enough to see the molecular structure of the cell organelles.

The contribution of electron microscopy to virology has been fundamental. For the first time it became possible to see virus particles not as dots at the limit of the resolution of the instrument, but as particles with a defined shape. The introduction of heavy metal shadowing made it possible to deduce the three-dimensional structure of the particles, and it became clear that each type of virus was characterized by elementary particles of characteristic form. It is interesting that it was found that many viruses had elementary particles whose shapes were those of the regular solids studied by geometers from the days of Plato; the cube, the tetrahedron, the dodecahedron, the icosahedron and the sphere. The way that these shapes are built up from assemblies of identical protein capsomeres is similar to the way in which the geometers construct the regular figures from identical right-angled triangles. Using ultra-thin sections and other techniques it has been possible to determine not only the relation between the protein capsid and the nucleic acid contained within it, but also the form taken by the nucleic acid itself.

The contributions of the light microscope to microbiological progress have not ceased. While the ultra-structure of the cell has been explored with the electron microscope, the development of phase contrast microscopy at the end of the nineteen-forties has made it possible to examine living, unstained cells at magnifications of up to ×1000. This has been the basis of many interesting and important studies on bacterial growth which have made use of a combination of phase contrast microscopy and time lapse cinematography. Such observations have made bacteriologists more aware of the importance of variation between the cells of a culture and have helped to elucidate the mode of action of certain antibiotics.

A recent technique to be applied to the examination of the structure of bacteria is that of autoradiography, where the radioactivity of an organelle is used to produce an image on film and so outline its shape and position within the cell. This technique was used brilliantly by Cairns in 1963 when he produced the first pictures that showed that the bacterial chromosome was in fact circular, and that it replicated in a way that had been predicted upon purely genetic evidence.

The scanning electron microscope which enables details of the surface structures of cells and colonies to be studied at magnifications of up to ×50 000 became commercially available in 1965 when the Cambridge Instrument Company put their 'Stereoscan' on the market. The first scanning electron microscope was built by Von Ardenne in 1938 but it was not until 1953 that the first machine with a short enough exposure time to enable cathode ray tube display was constructed by McMullen in Professor C. W. Oatley's Laboratory in Cambridge. Since 1967 when Gray published his Stereoscan pictures of soil microorganisms, an increasing number of papers on the application of this technique to microbiology have appeared.

REFERENCES

1677 Van Leeuwenhoek, A. *Phil. Trans. R. Soc.*, 13th yr, no. **133**, 821–31.

1849 Goeppert, H. R. & Cohn, F. Ueber die Rotation des Zellinhaltes in *Nitella flexilis. Bot. Ztg.*, **7**, 665–73, 681–91, 697–705, 713–19.

1869 Hoffman, H. Ueber Bakterien. *Bot. Ztg.*, **17**, 265–72.

1876 Weigert, C. Uber eine Mykose bei einem neugebornen kinde. *Jahresber. d. schlesischen Ges. f. vaterland. Cultur Breslau*, **53**, 229–30.

1876 Koch, R. Untersuchungen über Bakterien. V. Die Aetiologie der Milzbrand Krankheit, begrandet auf Entwicklungs geschichte des *Bacillus anthracis. Beitr. Biol. Pfl.*, **2**, 277–308.

1877 Koch, R. Verfahren zur Untersuchung zum Conserviren und Photographiren der Bacterien. *Beitr. Biol. Pfl.*, **2**, 399–434.

1881 Ehrlich, P. Ueber das Methylenblau und seine klinische-bacteriosco-pische Verwerthung. *Z. klin. Med.*, **2**, 710–13.

1882 Koch, R. Die Aetiologie der Tuberkulose. *Berl. klin. Wochenschr.*, **19**, 221–30.

1882 Ziehl, F. Zur Farbung des Tuberkelbacillus. *Dtsch. med. Wochenschr.*, **8**, 451–2.

1884 Gram, F. C. Ueber die isolirte Farbung der Schizomyceten in Schnitt und Trockenpräparaten. *Fortschr. Med.*, **2**, 185–9.

1885 Neelsen, F. See Johne, A. (1885). *Fortschr. Med.*, **3**, 198–202, (footnote p. 200).

1890 Loeffler, F. Weitere Untersuchungen über die Beizung und Farbung der Geisseln bei den Bakterien. *Zentralbl. Bakt.*, **7**, 625–39.

1894 Fischer, A. *Untersuchungen über Bakterien.* Berlin.

1919 Barnard, J. E. The limitations of microscopy. *J. R. microsc. Soc.*, 1–13.

1934 Marton, L. La microscopie electronique des objects biologiques. *Bull. Acad. Belg., Cl. Sci.*, **20**, 439–66.

1938 Von Ardenne, M. Das Elektronen-Rastermikroscop. *Z. Physik*, **109**, 553–72.

1941 Mudd, S. & Lackman, D. B. Bacterial morphology as shown by the electron microscope. *J. Bact.*, **41**, 415–20.

1942 Robinow, C. F. A study of the nuclear apparatus of bacteria. *Proc. R. Soc.*, Ser. B, **130**, 299–324.

1957 Dougherty, E. C. Neologisms needed for structure of primitive organisms. I. Types of nuclei. *J. Protozool.*, **4** (Supplement), 14.

1963 Cairns, J. The chromosome of *Escherichia coli. Cold Spring Harb. Symp. quant. Biol.*, **28**, 43–6.

1967 Gray, T. R. G. Stereoscan electron microscopy of soil micro-organisms. *Science*, **155**, 1668–70.

3
Artificial culture media

In 1860 Louis Pasteur published the first semi-synthetic medium designed for the growth of bacteria. This medium consisted of ammonium salts, yeast ash and candy sugar. Thus in the very first synthetic medium we can see the components that are required for bacterial growth; a nitrogen source (the ammonium salt), a carbon source (the candy sugar) and various vitamins and trace elements (the yeast ash). Prior to this bacteria had been grown in meat broths of uncertain constitution by various workers from Spallanzani onwards, but always the aim of their experiments had been to prove or disprove the theory of spontaneous generation and so the composition of the medium did not appear important to them. Pasteur's fluid on the other hand was developed to grow bacteria and yeasts under controlled conditions and in pure culture. The spontaneous generation controversy was ignored and it was assumed that only bacteria or othe micro-organisms that got into the culture medium either by intent or by chance would grow. The medium was devised by Pasteur to facilitate the work that he was engaged upon in the field of alcoholic fermentation and its 'diseases', and was in fact a simplified version of the must or wort that is the fermentable material used in the preparation of wine or beer.

In 1872, Ferdinand Cohn published the recipe for a similar medium that he had devised. This consisted of a basal medium containing the salts and the yeast ash, to which various sugars could be added at will. Cohn's medium is obviously more versatile than Pasteur's. It had been made up with the object of facilitating studies on the substrates which could and could not be fermented by various bacteria and fungi. A similar medium made up of a basal component and added sugars is still used for the same purpose today. The yeast ash has been replaced by mixtures of known salts and vitamins, but otherwise there has been no change.

All these early media were liquid, and the problems of obtaining pure cultures using such media were immense. Lord Lister gives an account of the methods employed: successive dilutions of the material under investigation were made until sterile samples were obtained. The penultimate tubes (in Lister's experiment sherry glasses were used) were taken to contain only a single organism and, therefore, to yield pure line cultures derived from the progeny of that single cell. Such cultures could then be transferred aseptically to other tubes of sterile culture medium to provide material for studies of the species of bacteria that had been isolated. Not only was the technique tedious but the results were uncertain and contamination was difficult to detect until a late stage in the investigation, depending for its recognition on direct microscopy of samples from the tubes. It is small wonder that with only such techniques available, the isolation of the causal micro-organisms of infectious diseases made little progress.

The introduction of solid media wrought a veritable revolution in bacteriology and made possible the flowering of the subject in the last two decades of the nineteenth century when the causal organisms of the majority of the bacterial infections were isolated and characterized. We owe this advance to one man, Robert Koch the great German bacteriologist (Fig. 3.1). Koch, who commenced his professional life as a country general practitioner, carried out microbiological research as a hobby and as we have seen was the first man to produce photomicrographs of bacteria as a part of his work on the aetiology of anthrax. The first solid medium employed by Koch was the cut surface of a boiled potato. The potato, having been sliced with a sterile (flamed) knife and protected from airborne contamination by placing it beneath a bell-jar, could be inoculated with material on the point of a needle. That material could then be spread over the surface so that at a certain point the dilution would leave separated single cells which would grow into colonies which would be pure, that is, descended from that single parent cell. It seems likely that this idea occurred to Koch as a result of observing the discrete colonies of pigmented micrococci which

Fig. 3.1. Robert Koch (1843–1910). From an original photograph in the Wellcome Historical Medical Museum. By courtesy of 'The Wellcome Trustees'.

grow on the face of cut potatoes exposed to the air. The cut potato technique was published by Koch in 1881, but already he had begun improving on it. In the same year Frederick Loeffler published the formula for his 'nutrient broth', a meat extract peptone medium designed to simulate the composition of the inflammatory exudates in which pathenogenic bacteria grow in the body. Koch set about developing a method for solidifying this medium. As an amateur photographer Koch was used to making up his own plates by coating sheets of plate glass with a solution of silver salts in gelatine. The first bacteriological culture plates were made by substituting nutrient broth for the silver salts and proceeding as if making a photographic plate. These plates were kept under bell-jars to protect them from aerial contamination and inoculated by stroking them with a needle charged with the material it was desired to culture, as in the case of the potato slices. These gelatine plates were very successful and enabled Koch to obtain individual colonies of organisms which were too fastidious to grow on the simple medium provided by the potato, but they had one great limitation; they could not be incubated at 37 °C, the body temperature of man and the optimum growth temperature of pathogenic bacteria, as the gelatine melted at that temperature. Koch announced his new culture medium at the International Medical Congress held in London at King's College in 1881, and it is said that after his communication Louis Pasteur went up to him and personally congratulated him on his discovery.

Later in the same year, Koch published details of another solid medium, inspissated serum. This medium is usually known as Loeffler's serum, but there is no doubt that the first person to publish the recipe was Robert Koch.

The problems arising from the low melting point of gelatine were solved when in 1882 agar was introduced as a gelling agent in media. This advance was due to Frau Hesse, the wife of one of Koch's assistants. Agar agar, a substance extracted from a Japanese seaweed, had been in use as a jam additive for many years. The Dutch had probably first brought it to Europe from their East Indian colonies. Like pectin, agar agar improved the setting quality of the jam. Its gels have the peculiar property of needing to be heated to 100 °C before they will melt, but once liquid, remain fluid down to a temperature of about 45 °C. Once agar had been tried out as a gelling agent for bacteriological media it was never abandoned, its peculiar melting point and gelling point, its poor nutritional value, its lack of inhibitory substances, all make it the ideal substance for the purpose of solidifying bacteriological media. Nutrient media solidified with agar were still, however, poured onto sheets of glass to form plates and had to be incubated and stored under bell-jars, so that a bacteriological laboratory of those days was littered with those cumbrous objects. Inspection of the culture plates could be made by looking through the glass, but subcultures could only be made by removing the bell-jar. The cumbrous nature of the apparatus and the bulk of it encouraged men to seek for some simpler container for media.

It was not until 1887 that the answer was found by Petri, another of Koch's assistants. The dishes which he described as slight modifications of the master's plates were an immediate success. The overlapping lid permitted a much clearer view of the colonies. When incubated with the surface of the medium downwards, both contamination and condensation were eliminated, and the dishes were much less bulky than the bell-jars and could be stacked one above the other in the incubator. Apart from the introduction of plastic instead of glass as the material for making the dishes, with the consequent advantage that they are disposable, no modification in design has been made since their invention.

In the same year, 1887, the first differential media were introduced. Chantemesse and Widal published an account of the use of glucose and lactose peptone water for the differentiation between *Escherichia coli* and *Salmonella typhi*. In these early days the acid production by the bacteria was measured by back titration with sodium hydroxide and gas analysis was by formal chemical means, so that such tests were tedious to perform.

The problem of telling the difference between the normal *E. coli* of the gut and the morphologically similar typhoid bacilli was made easier when, in 1889, the Japanese bacteriologist Kitasato developed a test for indole production, a property of *E. coli* but not *Sal. typhi* when they are grown in suitable peptone solutions.

It had become clear that the presence or absence of gas production during the fermentation of carbohydrates was an important characteristic for the classification of the pathogenic enterobacteria. The American bacteriologist Theobold Smith produced the first apparatus for the visual observation of gas production in 1890. His apparatus had a side arm in which the gas collected. It was efficient but the shape made it very liable to breakage during washing up. It was nevertheless a real advance on the previous methods.

In 1892, Wurtz of Paris introduced the use of indicators incorporated in the medium as a method of detecting acid production, thereby eliminating the need for time-consuming titrations.

In an article in the *British Medical Journal* of 1898, Durham, the British bacteriologist then working in Cambridge, published an account of a simple gas tube that could be placed within the test-tube containing the fermentable medium. These 'Durham's tubes' proved so much more convenient than Theobald Smith's apparatus that they immediately replaced it and are used routinely to this day.

The first and most famous solid differential medium was described by MacConkey (Fig. 3.2) in 1900 in *The Lancet*. The story of the development of this medium is of some interest. The first formula was for a bile salt litmus medium which contained

Fig. 3.2. A. T. MacConkey (1861–1931).
From *J. Path. Bact.* (1931), **34**, facing p. 698.

glycocholate, lactose and litmus, and which was to be incubated
at 22 °C. A year later MacConkey published a new formulation
of the medium in the *Zentralblatt für Bakteriologie*. The modified
medium contained taurocholate in place of glycocholate and was
to be incubated at 42 °C. Finally, in 1905, MacConkey published
a third formula for his medium, in which neutral red was
substituted for litmus as indicator. It is this last formula which we
use today for making MacConkey's medium. In fact, the use of
neutral red as an indicator in bile salt lactose medium was first
suggested in 1902 by Grunbaum and Hume in a paper in the
British Medical Journal.

About the turn of the century, bacteriologists began to
investigate the effect of the synthetic dyestuffs upon the growth
of bacteria in culture media. A systematic investigation of
various dyes; crystal violet, malachite green, brilliant green etc.,
was made by Conradi of Halle. As a result of this work Conradi

Fig. 3.3. M. W. Beijerinck (1851–1931) at seventy. Source unknown. By courtesy of 'The Wellcome Trustees'.

and Drigalski developed their medium for the selective cultivation of typhoid bacilli from faeces. This medium contained 1:100 000 crystal violet as an inhibitor of *E. coli*, peptone, lactose, and litmus as indicator.

In 1903, the Japanese bacteriologist Endo published another selective medium for the isolation of typhoid bacilli, this one depending upon the addition of basic fuchsin to a peptone lactose medium.

Malachite green was used in a similar way to suppress the growth of *E. coli* and permit the growth of *S. typhi* in a medium published by Loeffler in 1906. Thus by 1906 the techniques available for the isolation of intestinal pathogens, particularly *S. typhi*, included differential media containing lactose and an indicator, and a variety of inhibitory selective media depending upon the addition of dyes to the plates.

The year 1906 saw also the publication of Bordet and

Gengou's blood potato medium for the cultivation of *Bordetella pertussis*.

From 1898 onwards Beijerinck (Fig. 3.3) had been developing his technique of 'enrichment culture' for the isolation of species of soil micro-organisms. The principle of the method was to enrich the culture with respect to the species that it was desired to study by providing a medium in which the sole carbon or nitrogen source was one that could be utilized only by that species. The method had led to the isolation of various nitrifying bacteria and to the discovery of the cellulolytic bacteria. The extension of this technique to the isolation of pathogenic species from the rich mixture of organisms found in the stools was a logical step and it is surprising that it took so many years to come to be used. The first need was to find substrates which could be used only by, or more efficiently by, the pathogenic organisms, salmonellae and shigellae. The foundations for further development were laid by the work of Conradi and his associates who in about 1912 conducted a series of experiments to study the effect of salts of some rarer elements upon the growth of various bacterial species. In that year they published observations upon the effect of tellurium salts. Browning's (1913) method of isolating typhoid bacilli by plating out after overnight cultivation of the specimen in brilliant green broth was an early attempt at enrichment culture applied to medical bacteriology, but it depended upon the use of an inhibitory agent rather than upon the use of a substrate that would be better used by the pathogens. The first paper giving an account of a true enrichment technique for the isolation of typhoid bacilli from faeces is that of Guth published in the *Zentralblatt für Bakteriologie* in 1916. He suggested the use of selenite broth as an enrichment medium, and refers to previous work on the effect of selenite on bacterial growth by Haendel of Saarbrucken, in whose laboratory he was working.

The war of 1914–18 caused a large number of bacteriologists to devote themselves to the problems of wound sepsis, and in particular to study the bacteriology of gas gangrene, a frequent complication of battle wounds. It was known that the cause of this type of wound infection was members of the genus *Clostridium*, as the organisms could be seen with the microscope in the necrotic tissues, but the cultivation of such strict anaerobes

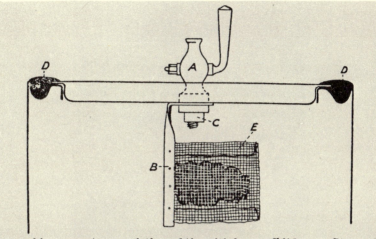

Anaerobic apparatus consisting of tin with lever-off lid. A = Stopcock;
B = brass bracket secured by C = collar of the tap; D = plasticine
filling in gutter; E = palladium capsule.

Fig. 3.4. A McIntosh and Fildes jar. From *The
Lancet* (1916), **i**, p. 768.

was difficult and their isolation from mixed flora such as those
encountered in soiled war wounds virtually impossible. The
available apparatus such as Veillons tubes were tedious to use
and liable to contamination. It was against this background that
McIntosh (Fig. 3.4) and Fildes invented their anaerobic jar in
which Petri dishes inoculated with swabs from wounds could be
incubated (Fig. 3.5).

The principle of the McIntosh and Fildes jar is well known.
After partial evacuation, hydrogen is added to the jar and then
an electric current is passed to heat up a platinized asbestos
catalyst which causes the residual oxygen to combine with the
hydrogen to produce water, thus leaving an anaerobic atmosphere within the jar. This apparatus enabled pure cultures of
clostridia to be grown from the most heavily contaminated
wounds and placed the study of the clostridia and other
anaerobic organisms upon a firm footing. The apparatus was
described for the first time in *The Lancet* in 1916. In the same
year Muriel Robertson published an account of her 'cooked
meat medium', another product of the work on the cultivation
of the anaerobic organisms of gas gangrene.

Fig. 3.5. J. McIntosh (1882–1948). From *J. Path. Bact.* (1949), **61**, facing p. 286.

In 1923, Muller of Liège described the use of tetrathionate broth for the selective culture of salmonellae and shigellae from stools. This medium is an enrichment medium depending on the fact that the two pathogenic genera, but not *E. coli*, are able to utilize tetrathionate as an alternative hydrogen acceptor to oxygen. Tetrathionate was the first rationally designed application of the principle of enrichment media to medical bacteriology.

The growth of public health bacteriology during the nineteen-twenties and nineteen-thirties led to the development of a number of selective media designed to facilitate the recognition of intestinal pathogens. Amongst the most important of these were Wilson and Blair's bismuth sulphate medium, the formula for which was first published in 1927, and the DCA medium formulation by Leifson of Johns Hopkins University in 1935. These inhibitory selective media were designed to be used in combination with the liquid enrichment media such as tetrathionate medium or the selenite medium introduced by Leifson in 1936.

Selective media for other organisms were also developed at this time. Sodium chloride agar for the isolation of staphylococci was introduced by Hill and White in 1929. The use of tellurite plates for the selective culture of *Corynebacterium diphtheriae* introduced by McLeod of Leeds in 1931 opened up a whole new area of the epidemiology of diphtheria and led to the recognition of the three types – gravis, mitis and intermedius.

Until the nineteen-thirties, media had been made by each laboratory for its own use and were dispensed in plates if solid and if liquid in test-tubes plugged with cotton wool. Different colours of cotton wool were used to indicate the medium in the tubes. The growth of the London County Council (LCC) hospital service involved setting up numerous laboratories, separated in space by many miles, but part of the same administrative unit. It became clear that it was not economic for each laboratory to make its own media and so the Director of Pathology for the LCC, Dr McCartney, set up a central medium kitchen from which all the laboratories in the service were supplied. It was obviously not practical to send media across London in cotton-wool-plugged test-tubes, and to get over this difficulty McCartney introduced the use of screw-capped bottles for containing media in convenient quantities for use at the bench. How he persuaded the United Glass manufacturers, who even at that time had a virtual monopoly of the manufacture of glass bottles in England, to make the special types of containers that he needed, how he arranged for the manufacture of the screw caps and liners for the containers, and how the central medium kitchen was organized, is to be found in the article that he published in *The Lancet* in 1933.

It was McCartney's central medium supply service that led later (1950) to the commercial development of dehydrated media. This was a logical extension of the practice in the food processing trade. Dehydration of foods had developed on a larger scale during the Second World War (1939–45) in order to save valuable shipping space, and the convenience of pre-cooked dehydrated dishes which had only to be reconstituted and heated caused these products to become popular in the postwar period. Appropriately enough, one of the first firms to offer dehydrated media for sale was a subsidiary of the food processors Oxo; Oxoid media are now well known all over the world. Dehydrated, prepared media have now almost completely replaced the old home-made media, and many of the younger bacteriologists have no experience of making nutrient broth from beef hearts in the traditional manner. The advantages of these dehydrated media are great. In the first place they save the time of technicians, and in the second they make a standard product available anywhere in the world. For critical work it is a great help to be able to purchase a sufficient quantity of a single batch of a medium so that conditions may be standardized throughout a series of experiments. The widespread use of these commercially prepared dehydrated media carries with it the danger that if anything goes wrong with a batch it will ruin the work of many laboratories, and this has led to more and more stringent testing of the quality of batches, both by chemical analysis and by biological methods. It is still true, however, that the data available on many commercial media are inadequate, and improvements are to be hoped for in this field.

During the 1939–45 war, interest in anaerobic wound infections came to the fore once again and led to the development of a very useful liquid medium for anaerobic bacteria, Brewer's thioglycollate medium, in which the Eh is kept at a low level by the incorporation of the reducing substance, thus enabling small inocula of strict anaerobes to grow.

In 1946 Christensen introduced his urea slopes for the identification of members of the genus *Proteus* and other urea-splitting organisms. These have become part of the routine method for differentiating between Enterobacteriaceae isolated from the stools.

In recent years a number of semi-synthetic media designed for specific purposes have been produced. Of particular interest are the media of Dubos and Kirschner for *Mycobacterium tuberculosis*, and the medium of Lacey for the isolation of *Bordetella pertussis*.

The future development of media is likely to be in the direction of the formulation of more selective media based upon our better understanding of bacterial nutrition and improvements in the standardization and batch testing of commercial dehydrated media.

REFERENCES

1860 Pasteur, L. Mémoire sur la fermentation alcoolique. *Ann. Chim. Phys.*, 3ᵉ sér., **58**, 323–46.

1872 Cohn, F. Untersuchungen über Bakterien. *Beitr. Biol. Pfl.*, **1**, 127–222.

1881 Koch, R. Zur Untersuchung von pathogenen Organismen. *Mitt. a. d. kaiserl. Gesundheitsamte*, **1**, 1–48.

1881 Loeffler, F. Zur Immunitätsfrage. *Mitt. a. d. kaiserl. Gesundheitsamte*, **1**, 134–87.

1882 Koch, R. Die Aetiologie der Tuberculose. *Berl. klin. Wochenschr.*, **19**, 221–30.

1883 Ehrlich, P. Einige Wörter über die Diazoreaktion. *Dtsch. med. Wochenschr.*, **9**, 549–50.

1887 Petri, R. J. Eine kleine Modifikation des Kochschen Plattenverfahrens. *Zentralbl. Bakt.*, **1**, 279–80.

1887 Chantemesse, A. & Widal, G. F. I. Le bacille typhique. *Gaz. hebd. Méd. Chir.*, 146–60.

1889 Kitasato, S. Die negative Indol-Reaktion der Typhusbacillen im Gegenstatz zu anderen ähnlichen Bacillenarten. *Z. Hyg. InfektKrankh.*, **7**, 515–20.

1890 Smith, T. Das Gahrungskolbchenin der Bakteriologie. *Zentralbl. Bakt.*, **7**, 502–6.

1890 Botkin, S. Eine einfache Methode zur Isolirung anaërobischer Bakterien. *Z. Hyg. InfektKrankh.*, **9**, 383–8.

1898 Durham, H. E. A simple method of demonstrating the production of gas by bacteria. *Br. med. J.*, **i**, 1387–9.

1900 MacConkey, A. Note on a new medium for the growth and differentiation of the bacillus *Coli communis* and the bacillus *Typhi abdominalis*. *Lancet*, **ii**, 20.

1901 MacConkey, A. Corrigendum et addendum. *Zentralbl. Bakt.*, **29**, 740.

1902 Conradi, H. & Drigalski, V. Ueber ein Verfahren zum Nachweiss der Typhusbacillen. *Z. Hyg.*, **39**, 283–300.

1902 Grunbaum, A. S. & Hume, E. H. Note on media for distinguishing *B. coli*, *B. typhosus* and related species. *Br. med. J.*, **i**, 1473–4.

1905 MacConkey, A. Lactose fermenting bacteria in faeces. *J. Hyg., Camb.*, **5**, 333–79.

1906 Bordet, J. & Gengou, O. Le microbe de la Coqueluche. *Ann. Inst. Pasteur*, **20**, 731–34.

1906 Loeffler, F. Der Kulturelle Nachweiss der Typhusbacillen in Faeces, Erde und Wasser mit Hilfe des Malachitgruns. *Dtsch. med. Wochenschr.*, **32**, 289–95.

1911 Holth, H. Untersuchungen über der Biologie des Abortusbacillus und die Immunitätsverhaltnisse des Infektiosen Abortus der Rinder. *Zentralbl. Infekt., Hautiere*, **10**, 207–73.

1913 Browning, C. H., Gilmour, W. & Mackie, J. T. The isolation of typhoid bacilli from faeces by means of Brilliant Green in fluid medium. *J. Hyg., Camb.*, **13**, 335–42.

1916 Guth, F. Selennahrboden für die selektive Zuchtung von Typhusbacillen. *Zentralbl. Bakt., Orig.*, **77**, 487–96.

1916 Robertson, M. Notes upon certain anaerobes isolated from wounds. *J. Path. Bact.*, **20**, 327–49.

1916 McIntosh, J. & Fildes, P. A new apparatus for the isolation and cultivation of anaerobic micro-organisms. *Lancet*, **i**, 768–70.

1923 Muller, L. Un nouveau milieu d'enrichessement pour la recherche du bacille typhique et des paratyphiques. *C. r. Séanc. Soc. Biol.*, **89**, 434–7.

1927 Wilson, W. J. & Blair, E. M. McV. Use of a glucose bismuth sulphate iron medium for the isolation of *B. typhosus* and *B. proteus. J. Hyg., Camb.*, **26**, 374–91.

1929 Hill, J. H. & White, E. C. Sodium chloride media for the separation of certain gram-positive cocci from gram-negative bacilli. *J. Bact.*, **18**, 43–57.

1931 Anderson, J. S., Happold, F. C., McLeod, J. W. & Thomson, J. G. On the existence of two forms of diphtheria bacillus – *B. diphtheria gravis* and *B. diphtheria mitis* – and a new medium for their differentiation and for the bacteriological diagnosis of diphtheria. *J. Path. Bact.*, **34**, 667–81.

1933 McCartney, J. E. Screw capped bottles in the preparation and storage of culture media. *Lancet*, **ii**, 433–6.

1935 Leifson, E. New culture medium based on sodium desoxycholate for the isolation of intestinal pathogens and for the enumeration of colon bacilli in milk and water. *J. Path. Bact.*, **40**, 581–9.

1936 Leifson, E. New selenite enrichment medium for the isolation of typhoid and paratyphoid (*Salmonella*) bacilli. *Am. J. Hyg.*, **24**, 423–32.

1940 Brewer, J. H. Clear liquid mediums for the 'aerobic' cultivation of anaerobes. *J. Am. med. Ass.*, **115**, 598–600.

1941 Hoyle, L. A tellurite blood–agar medium for the rapid diagnosis of diphtheria. *Lancet*, **i**, 175–6.

1946 Dubos, R. J. & Davis, B. D. Factors affecting the growth of the tubercle bacilli in liquid medium. *J. exp. Med.*, **83**, 409–23.

1946 Christensen, W. B. Urea decomposition as a means of differentiating proteus and paracolon cultures from each other and from salmonella and shigella types. *J. Bact.*, **52**, 461–6.

1954 Lacey, B. A new selective medium for *Haemophilus pertussis*, containing a diamidine, sodium fluoride and penicillin. *J. Hyg., Camb.*, **52**, 273–303.

4

Sterilization

Sterility of glassware and media is essential for the practice of microbiology. The first methods were developed during the controversy concerning spontaneous generation in the eighteenth century, but at that time it was not realized that bacterial spores were capable of withstanding boiling for periods of up to several hours, so there were many failures, and these were interpreted by the advocates of spontaneous generation as evidence in favour of their hypothesis.

From the earliest times boiling and flaming of glassware were used. Spallanzani in his classical work refuting the views of Needham and Buffon, published in 1765, boiled infusions for various periods of time in order to show that they would not undergo fermentation or putrefaction after this unless airborne microbes were admitted to the vessels in which they were kept. The later experiments of Schwann in 1836 and 1837 designed to settle the question of spontaneous generation also depended upon showing that infusions that had been boiled for some minutes could be preserved indefinitely so long as they were not contaminated with raw air. As it had been suggested that fermentation was a chemical process initiated by the presence of oxygen, it was necessary to show that it was not the air itself that caused the process, but the organisms suspended in it. It was to settle this point that Schwann designed his famous apparatus for admitting 'sterile' air to the infusions by passing the air through sulphuric acid or red-hot tubes. The results of these workers were always equivocal because they did not appreciate the existence of bacterial spores in their infusions.

These early microbiologists had a good deal of craft technique to draw upon. The autoclave had been in use for over one hundred years. In 1681 Denys Papin had published an account

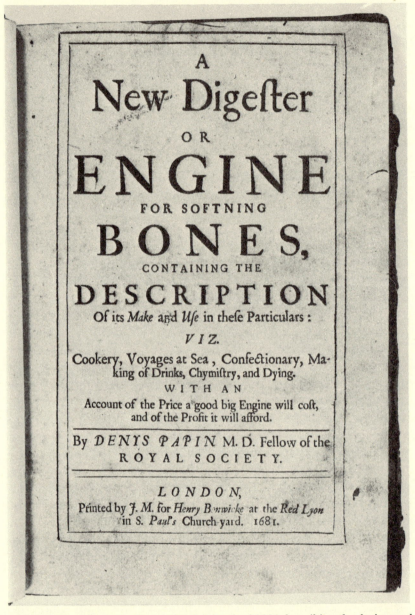

A

New Digester

OR

ENGINE

FOR SOFTNING

BONES,

CONTAINING THE

DESCRIPTION

Of its *Make* and *Use* in thefe Particulars:

VIZ.

Cookery, Voyages at Sea, Confectionary, Making of Drinks, Chymiftry, and Dying.

WITH AN

Account of the Price a good big Engine will coft, and of the Profit it will afford.

By *DENYS PAPIN* M. D. Fellow of the ROYAL SOCIETY.

LONDON,

Printed by *J. M.* for *Henry Bonwicke* at the *Red Lyon* in S. *Paul's* Church-yard. 1681.

Fig. 4.1. The title page of Denys Papin's book of 1681 describing the design and use of 'A new digester or engine for softning bones', which is in effect an autoclave.

of 'A new digester or engine for softening bones'. Containing the description of its make and use in 'cookery...chymistry and dying, with an account of the price a good big engine will cost and of the profit it will afford' (Fig. 4.1). Such engines were in common use in the glue making and food processing trades and were in fact made use of by early nineteenth century microbiologists from time to time. In 1801 Nicholas Appert, a French distiller and confectioner, published in Paris a little book entitled *The Art of Preserving Animal and Vegetable Substances for Several Years.* Appert's technique was to cork the substances up tightly in stout glass jars and then immerse them in boiling water for several hours. His preserved material was examined by some of the most eminent scientists of the day and found to be uncorrupted after many years. It was during the early years of the nineteenth century that the canning of meat and other products was first done on a commercial scale in England.

During his controversy with Pouchet concerning the possibility of spontaneous generation, which was at its height between 1858 and 1864, Pasteur made a number of important advances in the technique of sterilization, and in particular demonstrated the superior efficiency of wet heat over dry heat in sterilization. In 1861 he reported the results of experiments in which *Penicillium* cultures were heated in sealed flasks in an oil bath, some cultures being dry and others wet. The results were that whilst the fungus was killed at 121 °C wet heat, it survived 120 °C dry heat.

Later in 1876 when trying to explain Bastian's observation that alkaline urine could not be sterilized by boiling for several minutes, Pasteur demonstrated that sterilization could always be achieved if the urine was heated to 115–120 °C under pressure. This seems to have been the source of our traditional temperature for the sterilization of media.

It was not until 1877 that a true understanding of the problems of sterilization was achieved. In that year Tyndall showed that bacteria must exist in two forms, one heat-labile and the other heat-resistant. Independently the German botanist Cohn described the endospores of the hay bacillus and noted their resistance to heat.

Tyndall had been concerned with the nature of the particulate matter in the air since 1870 when he first used his technique of examining the air in a closed container by means of a beam of light passing across it, and showed that while at first the beam was clearly visible because of the myriads of particles suspended in the air, after a few days the particles settled out and the interior of the box appeared black even when a strong beam of light was passing across it. Tyndall, working at the Royal Institution, constructed a box in which it was possible to boil tubes of broth and hay infusion and then expose them to the air of the box. In a series of elegant experiments he showed that whenever the light beam was visible within the box there was bacterial growth in the boiled tubes, and whenever the particles had settled so that the interior of the box appeared black when a beam was shone through it, the tubes of media then exposed remained sterile for many weeks. Tyndall was thus able to show a correlation between the presence of the particulate matter in the air and the putrefaction or fermentation of the material exposed to it. When he came to repeat the experiments a few months later Tyndall found to his horror that the tubes in the dust-free air were now showing growth in the majority of cases, and the work seemed not to be reproducible. He shifted his laboratory to Kew, and there he was able to repeat the original experiments perfectly successfully. On his return to his laboratory in Albermarle Street, however, the tubes in the dust-free chambers once again showed growth. Tyndall then had a new laboratory built on the roof of the Royal Institution and when this had been thoroughly disinfected with phenol solution he set to work to repeat the experiments there. The results were in complete accord with the initial work in the Royal Institution and that at Kew. It was now clear that the cause of the failure must be due to something that had got into the downstairs laboratory between the first and second series of experiments. Investigations on the spot traced it to a bale of hay which had been brought into the laboratory to make hay infusion and had been left in a corner ever since. From the results of his experiments Tyndall was forced to the conclusion that bacteria could exist in two forms, one easily killed by a few minutes boiling and the other so resistant to heat that boiling for several hours was insufficient to cause death.

The contribution of Ferdinand Cohn was to give morphological identity to these heat-resistant forms and prove that they

Fig. 4.2. C. Chamberland (1851–1908). From the slide collection of the Department of Bacteriology and Virology, University of Manchester. Source unknown.

were the endospores found in the genus *Bacillus* and the genus *Clostridium*.

In 1881 Koch and his associates published two papers in which the minimum times and temperatures needed to achieve sterility of various materials by both hot air and steam were established. With regard to hot air they showed that it was necessary to achieve a temperature of at least 160 °C and to hold that temperature for one hour. Koch, Gaffky and Loeffler's paper entitled 'Observations on the effectiveness of hot steam for disinfection' in which quantitative work upon the temperature attained within rolls of fabric and the effect of load size upon the heat-up time of the material in the autoclave is reported, is now a classic. They showed amongst other things that after 30 minutes at 120 °C in the chamber the temperature at the centre of a roll of fabric was only 80 °C. These two papers laid the foundation of the practice of sterilization both in the laboratory and in the operating theatre.

The first laboratory autoclaves to be manufactured commercially were made by the French firm Weisnegg in 1884 to the design of Pasteur's colleague Chamberland (Fig. 4.2). Autoclaves appear to have been in use for culinary purposes in France for some time before this date. The name means a self-locking stew pan with a steam-tight lid; in fact, a pressure cooker. Chamberland's achievement was to design larger vessels for laboratory use and to persuade an engineering firm to manufacture them for the market. This was not Chamberland's only contribution to the practice of sterilization: he was also the designer of one of the first commercially available bacteriological filters, the so-called Chamberland's candles.

Chamberland's autoclaves were simply enlarged pressure cookers with a perforated plate at the bottom to keep the water away from the material that was to be sterilized. Displacement of air was upwards and therefore very inefficient and there was no means of drying the sterilized materials. So long as it was only bacteriological media that were being dealt with this was not a disadvantage, but for surgical dressings or any other fabrics it was a grave problem often leading to recontamination after removal from the chamber.

By 1898 the German engineers were producing autoclaves designed for hospital use with a venturi attachment which enabled a terminal vacuum to be drawn to dry the sterilized dressings, and also with a steam jacket to prevent condensation on the walls of the chamber (Fig. 4.3).

Although Frosch and Clarenbach had discussed the advisability of having downward displacement of the air in autoclaves in a paper published in 1890, it was not until 1916 that the first downward displacement machines were built, and even thirty years after that there were still many models on the market that did not have this feature. Why downward displacement took so long to be accepted as a necessary feature of autoclave design is something of a mystery. One has only to look at a boiling kettle to see that steam is lighter than air, and yet for decade after decade reputable instrument makers produced machines in which the displacement was upwards and supposedly intelligent bacteriologists and surgeons accepted them. It is a curious example of a mass blind spot.

Even with a steam jacket a certain amount of condensation

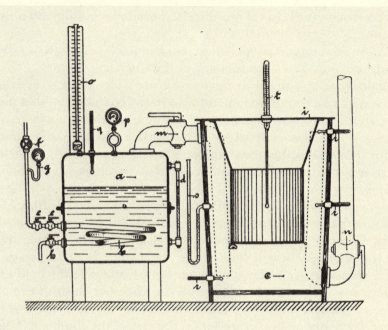

Fig. 4.3. A nineteenth-century autoclave. From *Z. Hyg. InfecktKrankh.* (1890), **9**, p. 188.

takes place within the chamber of an autoclave and upon the objects that are being sterilized in it. Such condensation not only makes the terminal drying of the materials more difficult but it impedes the penetration of dry steam into the middle of the load. The solution to this problem was found in the balanced pressure steam trap; a device placed in the chamber drain, which allowed the air and condensation water to pass out but held back the steam.

In essence the balanced pressure steam trap consists of a metal bellows, so set that when it is at a temperature above the boiling point of water the expansion of the air inside it forces the valve into its seating thus closing the trap: when, however, it comes into contact with any vapour or fluid that is below this temperature the air within the bellows contracts and the valve is withdrawn from its seating, so opening the trap and allowing the material to be expelled from the chamber. Such balanced pressure steam traps were being fitted to autoclaves made in

America and Germany in the early nineteen-twenties, but they were not manufactured nor fitted to autoclaves in England until 1932.

The next important development in autoclave design was the introduction of a thermometer in the discharge channel. This again was an American device. Before 1933 autoclaves had always had a thermometer installed in the chamber, so that the operator could note the temperature attained there. As we have seen it had been known since the work of Koch, Gaffky and Loeffler in 1881 that the chamber temperature was not a reliable guide to the temperature inside a load of dressings, which might be 40 deg C less. It was also known that the temperature within a loaded chamber was not uniform, some parts being 10–20 deg C hotter than others. Investigations showed that the coolest part of the chamber was always the chamber drain, therefore by placing a thermometer there it could always be assumed that the rest of the chamber would be at a higher temperature. If the holding period was not commenced until the chamber drain thermometer showed the desired temperature one could be sure that it had been attained or exceeded in the rest of the chamber. This increased the reliability of the sterilization, although of course it did not prevent nor detect the occurrence of regions at a lower temperature within packs.

To guard against the presence of these regions of sub-optimal temperature within drums and packs, a chemical indicator which could be placed in the centre of the packs of dressings was introduced. These are the little ampoules known as Browne's tubes which were invented in Britain by Dr A. W. Chapman, a lecturer in the Chemistry Department of Sheffield University and later Registrar at the University in 1925. An improved version was patented by him in 1937 and has been marketed by Albert Browne Ltd of Leicester since 1938. Since their first introduction many improvements have been made to these devices and now they are available as different types formulated to change colour from red through yellow to green after exposure to various defined temperatures and times – 121 °C for 15 minutes, 110 °C for 10 minutes and so on. The introduction of these indicators into drums or packs of dressings gives a visible check on the efficacy of the sterilization procedure when the drum is opened and before the contents are used, and has contributed much to the safety of hospital sterilization procedures.

During the twenty-five years from 1933 to 1958 no major change was made in the design of autoclaves. In that year a new era opened with the introduction by Knox and Penikett of pre-vacuum high-pressure sterilizers. These machines, which incorporated a mechanical pump to draw a vacuum of about 20 inches of mercury before the steam was admitted to the chamber and which operated at pressures of about 35 pounds per square inch, were claimed to have two great advantages. In the first place the penetration of the steam into the interstices of the fabrics was facilitated by the pre-vacuum and so more reliable sterilization was achieved, and in the second place the time taken to complete a sterilization cycle was cut from about forty-five minutes to fifteen minutes. Not only was the holding time cut to three or four minutes because of the higher temperature, but the times taken to heat up and to dry the load were cut dramatically by the use of the high vacuum.

A logical development of the pre-vacuum system was the pulsed vacuum system in which after the steam had been admitted to the chamber following the initial pre-vacuum a second and, in some cases, a third vacuum was drawn and steam admitted to the chamber. This system produced a progressive dilution of the residual air and could be relied upon to give almost pure steam in the chamber. It was first advocated in an article in the *Journal of Pharmacy and Pharmacology* by Wilkinson and Peacock in 1961. Trials conducted by Darmady and his associates led them by 1964 to the view that the simple single pre-vacuum technique was not sufficient and that a double vacuum should be drawn if the residual air was to be removed from loads of dressings before sterilization was commenced.

Two other recent developments in sterilization techniques are important. Both of them have developed to deal with the problems posed by the increased use of plastics in medicine. Plastic disposable syringes which are being used on an ever-larger scale are sterilized with gamma rays generated by cobalt-60 'bombs'. This method has only become possible as a result of the development of atomic energy and the commercial production of radioactive isotopes. The plant is very expensive and so the method is only economic when applied on a large scale. Up to the present it has been used almost entirely by the commercial firms that supply the disposable syringes, but there would seem to be a place for it within the National Health Service if regional central sterile units were to be set up, as it is capable of sterilizing almost all types of material. The second technique is gaseous sterilization, using ethylene oxide. This has been developed over the last ten years or so to deal in particular with the various prostheses increasingly inserted into the tissues of patients by surgeons, and the apparatus such as renal dialysis equipment that has to come in contact with the patient's bloodstream but is made of material which will not stand up to hot air sterilization or boiling.

A special problem is now posed by the sterilization of space probes and manned space vehicles. The complicated geometry and the multifarious substances used in the manufacture of these machines make them particularly difficult to deal with. Ethylene oxide has been used to sterilize the interior of unmanned probes (the exterior does not present a problem as it is heat-sterilized whilst passing through the earth's atmosphere), but manned vehicles present many as yet unsolved problems. It is unlikely that the present techniques such as those used on the 1969 moon flight are efficient enough to prevent either contamination of the surface of extra-terrestrial bodies visited by astronauts or the contamination of earth by extra-terrestrial microbes brought back by astronauts.

REFERENCES

1681 Papin, Denys. *A new digester or engine for softning bones, containing the description of its make and use in these particulars: cookery, voyages at sea, confectionary, making of drinks, chymistry and dying, with an account of the price a good big engine will cost and of the profit it will afford.*

1876 Pasteur, L. De l'influence de la temperature sur la fécondité des spores de mucédinées. *C. r. Acad. Sci.*, **83**, 377–8.

1876 Koch, R. Untersuchungen über Bakterien. V. Die Aetiologie der Milzbrand Krankheit, begrandet auf Entwicklungsgeschichte des *Bacillus anthracis. Beitr. Biol. Pfl.*, **2**, 277–308.

1877 Tyndall, J. Further researches on the deportment and vital persistence of putrefactive and infective organisms from a physical point of view. *Phil. Trans. R. Soc.*, **167**, 149–206.

1877 Cohn, F. Untersuchungen über Bakterien. IV. Beiträge zur Biologie der Bacillen. *Beitr. Biol. Pfl.*, **2**, 249–76.

1881 Koch, R. & Wolffhugel, G. Untersuchungen über die Desinfektion mit heisser Luft. *Mitt. a. d. kaiserl. Gesundheitsamte*, **1**, 301–21.

1881 Koch, R., Gaffky, G. & Loeffler, F. Versuche über die Verwerthbarkeit heisser Wasserdampfe zu Desinfektionszwecken. *Mitth. a. d. kaiserl. Gesundheitsamte*, **1**, 322–441.

1890 Frosch, P. & Clarenbach, A. Uber das Verhalten des Wasserdampfes im Desinfektionsapparate. *Z. Hyg. InfecktKrankh.*, **9**, 183–217.

1925 Hall, R. J. B. & Chapman, A. W. A study of the efficiency of sterilization of dressings. *Br. Med. J.*, **i**, 1119–20.

1936 Shrader, H. & Bossert, E. US Patent No. 2 637 439.

1949 Phillips, C. R. & Kaye, S. The sterilizing action of gaseous ethylene oxide: I, II, III, IV. *Am. J. Hyg.*, **50**, 270–305.

1950 Wilson, A. T. & Bruno, P. The sterilization of bacteriological media and other fluids with ethylene oxide. *J. exp. Med.*, **91**, 449–58.

1958 Knox, R. & Penikett, E. J. K. Influence of initial vacuum on steam sterilization of dresings. *Br. med. J.*, **i**, 680–2.

1961 Wilkinson, G. R. & Peacock, F. G. The removal of air during autoclave sterilisation of fabrics using low pressure steam. *J. Pharm. Pharmacol.*, **13**, (Supplement 67).

1961 Bowie, J. H. *On Recent Developments in the Sterilization of Surgical Materials*. Pp. 109–42. London: Pharmaceutical Press.

1964 Darmady, E. N., Drewett, S. E. & Hughes, K. E. A. Survey on pre-vacuum high-pressure steam sterilizers. *J. clin. Path.*, **17**, 126–9.

5

Chemotherapy

Chemotherapy in the modern sense, that is the synthesis of drugs which have a specific effect on the parasites, killing them or preventing them from multiplying in the body while producing little or no toxic effects upon the host, is the invention of one man, Paul Ehrlich (Fig. 5.1). Ehrlich, a German Jew of polymathic chemical and biological knowledge and incredible industry, developed the concept of the 'magic bullet' which would wipe out the parasites from the body of the host following a single *dosa sterilisa magna*, had the vision to apply what he had learned about specific staining of cells to the problem of synthesizing chemotherapeutic agents, and with the co-operation of the German chemical industry, produced and put on the market the first effective chemotherapeutic agents. His theories of specific receptors and differential affinities dominated the field for twenty years after his death and were fruitful in producing such agents as mepercrine, the first synthetic malarial prophylactic and prontosil red, the parent compound of the sulphonamides.

In additon to his work on chemotherapy, Paul Ehrlich was the inventor of the differential stain for white blood cells from which the modern Romanovsky methods have been developed, and was the initiator of quantitive methods in immunology when he was in charge of the German State Serum Institute. His work on the nature and origin of antibodies, the famous 'side chain theory', dominated thinking in this field for a generation and will be discussed at length in the chapter on theories of antibody formation. There have been few medical scientists who have made major contributions in so many different fields.

There had been previous use of specific antimicrobial agents in medicine, but these had been empirical in origin, and as they

Fig. 5.1. Paul Ehrlich (1854–1915). From an original photograph by Eduard Blum. By courtesy of 'The Wellcome Trustees'.

had been discovered before the germ theory of disease was accepted, their mode of action had been misunderstood. The use of ipecacuanha for the treatment of amoebic dysentery, which was advocated by Chinese physicians as long ago as 500 BC depended upon the specific action of emetine. The use of cinchona bark for the treatment of malarious fevers, introduced into Europe by returning Jesuit missionaries in the early seventeenth century, but in use in Peru for we know not how long before that, depended upon the quinine content of the bark. The treatment of syphilis with mercury, introduced by Paracelsus at the end of the sixteenth century was another example of the use of a specific chemotherapeutic agent, but was explained in alchemical not biological terms.

A curious feature of the history of chemotherapy is the long gap that seems always to have occurred between the synthesis of a compound by the organic chemists and its first use therapeutically. Arsenilic acid was first synthesized by the French chemist Bechamp in 1863, but it was not until 1902 that Ehrlich tested its derivative atoxyl in the treatment of experimental trypanosomiasis, a lag period of thirty-nine years. Again, sulphonamide was first synthesized by Gelmo in 1908, but it was not until 1935 that Domagk tested the sulphonamide derivative prontosil red in the treatment of experimental pneumococcal infections, a period of twenty-eight years.

There can be no doubt that Paul Ehrlich developed his theories concerning chemotherapy from the basis of his work on differential staining of cells in tissues. His earliest paper, published while he was still a student in 1877, was on the use of aniline dyes in microscopy. In 1880 he published an account of his famous tri-acid stain for the differential staining of leucocytes. It was in this paper that he coined the terms acidophile leucocyte, basophile leucocyte and neutrophile leucocyte.

In 1885 Ehrlich published a monograph entitled *The Oxygen Requirement of the Organism*. This little book added another thread to his thinking. For the first time dyes were used as oxidation/reduction indicators to demonstrate the conditions within cells that were actively metabolizing. Amongst other dyes Ehrlich used methylene blue as a vital stain and noted that it was taken up differentially by various cells. It was this observation which led him a few years later to the first of his ventures in chemotherapy.

In 1891 Ehrlich published a paper describing the use of methylene blue for the treatment of malaria. He had noted that the dye stained malaria parasites in blood films much more densely than it stained the blood cells and so was led to try out the effect of the dye when given intravenously to patients with malaria. The drug proved effective, clearing the bloodstream of parasites and producing clinical cure. It was, however, no better than quinine and as it had the disadvantage of turning the patient's urine blue, it never came into general use. Although this treatment of malaria is now only a historical curiosity it was nevertheless a very important step in the development of chemotherapy. For the first time a synthetic drug had been used in a rational way to cure the patient by killing the parasites: perhaps even more important the effect had been successfully predicted from the action of the drug on the parasites and host cell *in vitro*. Many of Ehrlich's later studies were to involve the twin themes of the differential uptake of dyes and the use of a test system which enabled him to assess the effect of the drug on the infection by microscopy of blood smears to detect parasitaemia.

By 1902 Ehrlich was devoting almost all his energies to the investigation of synthetic compounds for chemotherapeutic activity. He concerned himself with a series of organic arsenicals, and his test system was experimental trypanosomiasis in mice. He was able to estimate the effect of the drugs on the infected mice by the simple method of cutting a small piece off the tail and making blood smears from the bleeding stump. Positive results showed as a disappearance of parasitaemia. The infection could be easily passaged from one mouse to another by injection of blood from an infected mouse into an uninfected one. The Japanese bacteriologist Shiga was at that time working with Ehrlich and in that year they published a paper in which they concluded that atoxyl (sodium arsenilate) was ineffective in the treatment of mouse trypanosomiasis. They were mistaken. Atoxyl is a very effective trypanocidal agent. What had happened was that they had been using, unknown to themselves, a strain of trypanosome that was naturally resistant to arsenicals.

In 1904 Ehrlich and Shiga announced their first successful chemotherapeutic agent – trypan red. This drug when tested using the mouse trypanosomiasis system cleared the blood of parasites rapidly, did not have any marked toxic effects on the mice, and appeared to effect a cure of the infection.

In 1905 two important things happened. The British bacteriologist Thomas re-investigated the effect of atoxyl in trypanosomiasis on the mouse, and, as he used a sensitive strain, reported that it was very efficient trypanocide indeed. This re-opened the investigation of the organic arsenicals based upon modifications to the arsenilic acid molecule.

The second momentous event was the discovery of the *Treponema pallidum*, the causative organism of syphilis, by Shaudin and Hoffman, and the transmission of the infection to chimpanzees. Shaudin and Hoffman believed that the organism they had discovered was a protozoan, and Ehrlich accepted their classification of the parasite. Thus the rectification of the error concerning atoxyl and the erroneous classification of the treponema led him to speculate as to the possible effect of his organic arsenicals upon the causative organism of syphilis. The treatment of syphilis at that time was limited to the inunction of mercury and the exhibition of potassium iodide, and very few cases were ever completely cured. The disease was widespread in Europe and cases of tertiary vascular syphilis and neurosyphilis were very common. It was not for nothing that syphilis was called the great mimic.

In 1906 Robert Koch first used atoxyl for the treatment of human cases of trypanosomiasis. The disease was at that time epidemic in East Africa where the Germans had colonial possessions, and it was there that the first therapeutic trials were carried out. As the disease was almost uniformly fatal when not treated, the results of the trials were held to be very encouraging even though the drug was only curative if given in the early pre-neurological stages of the disease and was not without toxic hazards, a proportion of the patients developing bilateral optic atrophy as a result of the treatment. Atoxyl continued to be used for the treatment of trypanosomiasis for some years until it was replaced by safer and more effective drugs.

The year 1906 also saw the opening of the George Speyer House, the first institute for experimental chemotherapy. The institute was built by the widow of George Speyer, a wealthy German manufacturing chemist, as a memorial to her husband. Paul Ehrlich was asked to become the first director, and accepted. He now had control of a well equipped institute with chemical and biological laboratories in plenty and very generous endowments for the purchase of equipment and materials and the salaries of assistants. The team of chemists and bacteriologists that he built up was to become world famous, and he himself became a legendary figure: the Geheimrat Professor with his eternal cigar and his encyclopaedic learning; his habit of drawing chemical formulae on the floor or on the walls to illustrate a point; his hoard of organic chemicals that must never be thrown away; and his absent-mindedness with regard to the affairs of ordinary life when engaged with a chemical or biological problem, was a beloved if rather awesome figure to all his staff and has, since his death, been taken as an archetypal figure for the medical students.

The discovery of the curative effect of salvarsan in syphilis came about in a rather curious way. According to Martha Marquardt, Ehrlich's secretary, the compound was first synthesized in Ehrlich's laboratory in 1907. It was tested for therapeutic

Fig. 5.2. Paul Ehrlich (1854–1915) and Sachachiro Hata (1873–1938). From a photogravure of *c.* 1910 in the Ehrlich Collection at the Wellcome Institute. By courtesy of 'The Wellcome Trustees'.

activity using the experimental mouse trypanosomiasis system and found to be inactive, and no further tests were carried out on the compound at that time.

In 1909 a young Japanese scientist, Hata, came to work in Ehrlich's laboratory (Fig. 5.2). Hata, working in Kitasato's laboratory, had been responsible for developing a system for the artificial transmission of *Treponema pallidum* in rabbits by means of intratesticular inoculation. Ehrlich and Kitasato were old friends, having worked together as young men in Koch's laboratory, and it would seem that Hata's visit was intended to be a two-way exchange of information. Hata was to instruct Ehrlich's colleagues in the technique of experimental transmission of syphilis in the rabbit, and Ehrlich was to give Hata training in experimental chemotherapy. As an initial project Hata was to examine the chemotherapeutic effect of a number of compounds whose efficiency or otherwise was already known, using the mouse trypanosomiasis system to screen them: a normal procedure for teaching a young worker a new technique. Amongst the compounds that he was given to test was No. 606 (salvarsan).

To everybody's surprise he reported that the compound was highly active against trypanosomal infection in mice. The experiments were repeated and the unexpected results were confirmed. It remains a mystery how Ehrlich failed to observe the effect in 1907. By now Hata had got his experimental rabbit syphilis system going in Ehrlich's laboratory, and naturally he started working through the organic arsenic compounds that were known to be active against the trypanosomes to see if they were also active against spirochaetes. In the control infected animals the testicle was swarming with spirochaetes which could easily be demonstrated in smears from the excised organ by dark ground microscopy: in animals treated with an intravenous dose of salvarsan the material was completely clear of spirochaetes, and when portions of the ground-up testicles were injected into the testicles of further rabbits no infection followed. These experiments were repeated again and again and always the result was the same: a single dose of salvarsan completely cured the experimental infection in the rabbits (Fig. 5.3). Extended toxicity trials were carried out which showed that the ratio of therapeutic:toxic dose was higher than for any of the other organic arsenicals tested. Finally, late in 1909 the drug was given its first trial on human patients suffering from syphilis. The results reported by Hata were dramatic. After a single intravenous injection, primary chancres healed completely and secondary rashes and snail-track ulcers disappeared in a few days. Spirochaetes could not be demonstrated in the healing lesions and so it appeared that the patients had been rendered

TREATED RABBIT No. VIII.

a

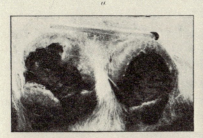

ON THE DAY OF TREATMENT.

b

FOUR DAYS AFTER TREATMENT.

c

EIGHT DAYS AFTER TREATMENT.

d

FIFTEEN DAYS AFTER TREATMENT.

e

TWENTY-TWO DAYS AFTER TREATMENT.

f

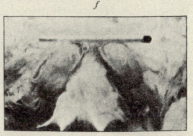

TWENTY-EIGHT DAYS AFTER TREATMENT.

Fig. 5.3. The experimental chemotherapy with Salvarsan of rabbits infected with syphilis. From P. Ehrlich & S. Hata, *The Experimental Chemotherapy of Spirilloses.* Berlin, Springer-Verlag (1910). (English translation, London, 1911.)

3-2

uninfective. The chemical section of the George Speyer House worked overtime to produce supplies of salvarsan for the clinical trials and to meet the enormous demand for the drug that was foreseen when the discovery of a cure for syphilis was made public. Later, of course, the drug would be made commercially by pharmaceutical firms, but for the first weeks or months all supplies had to come from Ehrlich's laboratory. It was fortunate that the institute had been planned on a large scale as it was possible to set aside laboratory space for what we should now call pilot scale production of the drug.

The first public announcement of the discovery was made by Ehrlich at the Medical Congress at Wiesbaden on 19 April 1910. When his communication was completed he received a standing ovation. The physicians present realized the immense importance of his discovery. A mass of human misery could now be done away with. One of the major plagues of the human race could be eradicated, and generations as yet unborn would bless the name of Ehrlich and remember as auspicious the day of his announcement.

In 1911 Morgenroth, who had been the head of the chemical section of Ehrlich's institute but now was director of his own laboratory in Berlin, discovered another type of chemotherapeutic agent. He was working on pneumococci and one of his assistants was working on the effect of quinine derivatives in malaria. He tried out the effect of one of these, optochin, upon the pneumococci in the test-tube and found it to be bactericidal in high dilution. He then tested optochin *in vivo* treating mice experimentally infected with pneumococci with the drug. The experimental pneumococcal infections were cured, but the toxic effects of the compound were too severe for it to be of any use in the treatment of human infections. Today, optochin is used only as a laboratory reagent to differentiate pneumococci from *Streptococcus viridans* which is insensitive to the action of the drug.

One of the problems that vexed early workers in the field of chemotherapy was how to determine whether a substance that was bacteriostatic in the test-tube would be effective in curing experimental infections in animals. Many of the compounds that were investigated proved to be quite ineffective *in vivo* even though they were active at high dilution *in vitro*. The animal experiments to prove this had to make use of quite large control and test groups and were tedious and expensive. In 1912, Sir Almroth Wright, Professor of Pathology at St Mary's Hospital in London, found a solution to this problem. He noted that while there was little or no association between the in-vitro and in-vivo results when the tests were made growing the bacteria in nutrient broth or peptone, if the tests were made in the presence of serum then the results were in agreement. No inhibition of growth in the test-tube in the presence of serum indicated a failure to cure experimental infections, while inhibition in the presence of serum usually indicated some therapeutic effect on the experimental animals.

Paul Ehrlich died full of years and honours in 1915, but the search for specific chemotherapeutic substances that he had initiated so brilliantly went on. All over the world his disciples applied the principles that he had laid down to the synthesis and testing of series of organic compounds. This work led to the discovery of a number of useful agents. The use of acriflavine as a wound disinfectant came as a result of the researches of Browning and his colleagues at Glasgow during the First World War. Morgenroth and his collaborators synthesized a number of quinoline derivatives with local antiseptic activity.

In 1932 Atebrin, the first synthetic drug for the prophylaxis of malaria was synthesized in the laboratories of the German chemical firm of Bayer. This substance, an acridine dye, can be thought of as a lineal descendant of methylene blue, shown by Erlich to be active against malaria so many years before. Certainly without Ehrlich's work as a starting point it is doubtful whether the compound would ever have been synthesized.

The discovery of the curative action of prontosil red in experimental pneumococcal infections by Domagk in 1935 was a piece of work in the classical Ehrlich tradition. Domagk had been at work testing the chemotherapeutic activity of various synthetic dyestuffs for some years. The dye prontosil red was found to be very active in suppressing the growth of pneumococci in the test-tube, and when tested upon groups of mice experimentally infected with pneumococci by intraperitoneal injection it was found to have a powerful curative action. As the dye was of very low toxicity it was given therapeutic trials in cases of lobar pneumonia in humans, and the results were dramatic, the death rate being reduced from about 30 per cent to under 10 per cent.

Later in the same year the French worker Tréfouël and his

colleagues studying the mode of action of prontosil red split the molecule into two parts, and found that the anti-bacterial activity resided not in the chromophore part but in the colourless sulphonamide part (sulphanilamide).

This discovery made it possible to progress to the synthesis of modified sulphanilamide molecules with different pharmacological properties and thus suited to the treatment of different types of infection. There are now dozens of sulphonamides available all made by modifying the parent molecule at the NH_2 end. In 1968, a new and still more powerful antibacterial agent, co-trimoxazole, was produced by combining a sulphonamide with trimethoprim, a folic acid antagonist, and thus blocking the synthetic pathway for folic acid at two places. As might be expected this combination substantially decreases the likelihood of resistant strains emerging.

It was not until 1940 that the mode of action of the sulphonamides was elucidated. In that year D. D. Woods and the veteran microbiologist Sir Paul Fildes, showed that these compounds acted by competing for an enzyme site with *para*-aminobenzoic acid (PABA) an essential nutritional requirement for many bacteria.

The action of the sulphonamides could be reversed by the addition of PABA to the medium and quantitative studies proved that the affinity of PABA for the enzyme site was about fifty times that of the sulphonamide molecule. This was the first time that the mode of action of an antibacterial compound had been explained at the molecular level and naturally high hopes were raised that it would lead to the synthesis of other antimetabolites which would act in a similar way by competitive inhibition on other synthetic pathways of bacteria. In the event the search for antimetabolites has not proved as easy as was expected and only a very few effective agents have been produced, and most of these have been of value in cancer therapy, not in the treatment of bacterial infections. The most probable reason for the paucity of results from this rational approach to antibacterial therapy was limited knowledge of the synthetic pathways and control mechanisms of both bacterial and mammalian cells. Many of the antimetabolites synthesized were useless, not because they did not inhibit the growth of bacterial cells, but because they blocked pathways common to the cells of both host and parasite, and were therefore too toxic for use. The recent development of co-trimoxazole mentioned above and of certain of the newer antiviral agents based on modified DNA suggest that we may now, with a much more sophisticated model of cell organization at the molecular level, be about to make rapid progress in the design of antimetabolites with differential actions upon host and parasite.

The synthesis of paludrine, a new and very efficient malarial suppressive, by Rose working in the ICI laboratories in 1946 suggested a whole group of new compounds that should be investigated, as well as revolutionizing the prospects for malaria control all over the world. A few years later in 1955, another potent synthetic antimalarial which had the advantage of needing to be taken only once a week was produced, this time in the research laboratories of Burroughs Wellcome and Company. This compound, pyrimethamine (marketed under the trade name of Daraprim) is active against both exo-erythrocytic and blood forms of the plasmodia. It is also active against *Toxoplasma gondii*.

In 1960 Beecham Laboratories announced the production of the first semi-synthetic penicillins. These compounds are made by so arranging the fermentation that the product is penicillinic acid, the core of all types of penicillin, harvesting this material and then inserting specific side groups by chemical means. It has proved possible by means of this approach to make penicillins which are acid-resistant, others resistant to the action of staphylococcal penicillinase, and others such as ampicillin which have an antibiotic spectrum as wide as the tetracyclines. So far this approach of using a micro-organism to synthesize the nucleus of an antibacterial compound and then modifying it at will by orthodox chemical means has not been applied to any other type of antibiotic.

Since 1960 the growing point of chemotherapeutic research has been the attempt to produce compounds that are effective against viruses. The problem is one of great difficulty for two reasons: in the first place viruses are so intimately involved with the metabolism of the host cells that it is almost impossible to block metabolic steps which are peculiar to the parasite; in the second place in most virus diseases the diagnosis, indeed the clinical illness, does not occur until the body is already saturated with virus. Thus treatment cannot be started until the infection is already at, or past, its peak of intensity. In spite of these difficulties some progress has been made. Two main lines are being explored: on the one hand it has sometimes proved possible to inhibit the enzymes which the virus uses to attach itself to the cells of the host; on the other hand there are a number of compounds that either stimulate the production of interferon in the host cells, as does Statolon, originally believed to be a fungal product but now shown to be a virus, or release interferon which is already pre-formed in the cells, as does the synthetic two-stranded RNA polyinosinic acid/polycytidylic acid, referred to as poly I/poly C. This approach of designing drugs which will mobilize the natural cellular defences would seem to hold out the greatest hope of progress in the new field of virus chemotherapy, but even if compounds active in this way and non-toxic to the host can be synthesized, and presently available compounds have various undesirable side-effects, producing a febrile response and damaging the foetus in pregnant animals to name only two, there will still be the problem of the late appearance of clinical illness in viral infections. One of the most effective antiviral compounds synthesized so far is a sulphone derivative, Marbaran, first used in 1963. It is of great value in the prophylaxis of smallpox, but has no effect upon the course of the disease in clinical cases.

REFERENCES

1885 Ehrlich, P. *Das Sauerstoff-Bedurfniss des Organismus.* Berlin: Hirschwald.

1891 Guttmann, P. & Ehrlich, P. Ueber die Wirkung des Methylenblau bei Malaria. *Berl. klin. Wochenschr.*, **28**, 953–6.

1905 Thomas, H. W. & Breinl, A. *Repeort on trypanosomes, trypanosomiasis and sleeping sickness.* London: University of Liverpool.

1910 Hata, S. *Verhandlungen des Congresses für Innere Medizin, Wiesbaden,* p. 225.

1911 Ehrlich, P. & Hata, S. *The Experimental Chemotherapy of Spirilloses,* pp. 64 and 70. London. (Originally published in German by Springer-Verlag, Berlin in 1910.)

1911 Morgenroth, J. & Levy, R. Chemotherapie der Pneumokokkenin-infektion. *Berl. klin. Wochenschr.*, **48**, 1979–83.

1913 Browning, C. H. & Gilmour, W. Bacterial action and chemical constitu-tion with special reference to basic benzol derivatives. *J. Path. Bact.*, **18**, 144–6.

1919 Browning, C. H., Gulbransen, R. & Kennaway, E. L. Hydrogen-ion concentration and antiseptic potency, with special reference to the action of acidine concentrations. *J. Path. Bact.*, **23**, 106–8.

1929 Browning, C. H. *et al.* The trypanocidal action of some derivatives of anil and styryl quinoline. *Proc. R. Soc.*, Ser. B, **105**, 99–111.

1935 Domagk, G. Ein Beitrag zur Chemotherapie der bakterillen Infektionen. *Dtsch. med Wochenschr.*, **61**, 250–3.

1935 Tréfouël, J. *et al.* Activité du P-aminophénylsulfamide sur les infections streptococciques expérimentales de la souris et du lapin. *C. r. Séanc. Soc. Biol.*, **120**, 756–8.

1946 Curd, F. H. S. & Rose, F. L. Synthetic antimalarials. X. Some aryl-diguanide ('-biguamide') derivatives. *J. Chem. Soc.*, 729–37.

1949 Marquardt, M. *Paul Ehrlich.* London: W. Heinemann.

1954 Marquardt, M. Paul Ehrlich – some reminiscences. *Br. med. J.*, **i**, 665–7.

1961 Herrmann, E. C. Plaque inhibition test for detection of specific inhibitors of D.N.A. containing viruses. *Proc. Soc. exp. Biol. Med.*, **107**, 142–5.

1963 Baurer, D. J., St Vincent, L., Kempe, C. H. & Downie, A. W. Prophylactic treatment of smallpox contacts with N-methylisatin-β-thiosemicarbazone (compound 33T57, Marboran). *Lancet*, **ii**, 494–6.

1966 Bryans, J. T. *et al.* 1-Adamantanamine hydrochloride prophylaxis for experimentally induced A/equine 2 influenza virus infection. *Nature, Lond.*, **212**, 1542–4.

1967 Marshall, W. J. S. Herpes simplex encephalitis treated with idoxuridine and external decompression. *Lancet*, **ii**, 579–80.

1969 Absher, M. *et al.* Toxic properties of a synthetic double-standed RNA. *Nature, Lond.*, **223**, 715–17.

1969 Park, J. H. & Baron, S. Herpetic keratoconjunctivitis: therapy with synthetic double-stranded RNA. *Science*, **162**, 811–13.

1970 Juel-Jensen, B. E. Severe generalised primary herpes treated with cytarabine. *Br. med. J.*, **ii**, 154–5.

1971 Galbraith, W. W. *et al.* Therapeutic effect of 1-adamantanamine hydrochloride in naturally occurring influenza A2/Hong Kong infection. *Lancet*, **ii**, 113–15.

1973 Fortuny, I. E. *et al.* Cytosine arabinoside in *Herpes zoster. Lancet*, **i**, 38.

6

Antibiotics

Antibiotics are regarded as one of the greatest triumphs of microbiology, but the tale of their discovery is one of the least edifying chapters in the history of the subject. Time and again microbial antagonisms have been recognized and reported, time and again the therapeutic significance of these observations has been discussed in the scientific press, compounds have been isolated and small-scale clinical trials have been conducted, and each time the clues have not been followed up, the observations have been ignored and forgotten, and all the work has had to be done again, ten, twenty and thirty years later.

When finally the isolation of effective antibiotics was achieved, it was done by hit-and-miss methods involving chance observations in the first place, and later the screening of thousands of species of soil micro-organisms from all over the world. The base empiricism of such a programme contrasts most unfavourably with the attempts at rational progress which characterize the development of chemotherapy. The mode of action of the antibiotics was studied after their discovery by microbial physiologists in no way connected with the search for such substances, and the site of action of many of the common antibiotics is still uncertain.

It is interesting to speculate upon the cause of this difference in approach. Did it spring from the state of development of the sciences of biology and chemistry, the former still only just emerging from the stage of classification and description, the latter equipped with a well-developed theoretical framework and capable of quantitative predictive statements? Was it an historical accident arising from the lack of any single scholar with a continuing programme of research in the field, comparable to Paul Ehrlich's long-continued systematic investigation of chemotherapeutic substances? Was it the result of attempts to apply the results of the research to practical medicine too soon?

The first published account of microbial antagonism is that of William Roberts in the *Philosophical Transactions of the Royal Society* in 1874. He noted that the growth of fungi often suppressed the growth of bacteria and goes on to say: 'I have repeatedly observed that liquids in which the *Penicillium glaucum* was growing luxuriantly could with difficulty be artificially infected with bacteria'. This observation which specifically mentions the effect of a *Penicillium* on the growth of bacteria and foreshadows the work of Fleming fifty years later, did not attract attention and was forgotten until it was discovered by medical historians within the last ten years. Two years later John Tyndall published an account of antagonism between a *Penicillium* and various bacteria in liquid cultures and noted that in the struggle between the mould and the bacteria the mould was usually successful. It appears that this was a completely independent observation on his part. Once again the observations were not followed up and were lost sight of for many years. The next year, 1877, saw the publication of a paper by Pasteur and Joubert in which they reported their observation that animals infected with cultures of anthrax bacilli that had become contaminated with saprophytic bacteria did not develop the disease. They noted, 'these facts may offer great hopes from a therapeutic point of view'. It might seem that the way forward was now clear, but once again the results were ignored by other microbiologists and slipped into the limbo of forgotten papers, to be recovered only after the discovery of penicillin and streptomycin. Another approach to the problem of antagonism between microbes was the work of Babes and Cornil published in 1885. They studied the reciprocal action that bacteria have one upon another, using amongst other techniques the cross streak method, now routinely in use for studies on both antibiotic effects and cross-feeding. They speculated that such studies might lead to therapeutic advances. Using the cross streak test Garré in 1887 showed that *Pseudomonas pyocyanea* produced a specific diffusible substance which was able to inhibit the growth of various bacteria including staphylococci.

Interest in these diffusible substances from *Pseudomonas pyocyanea* led to a spate of papers on the subject of the antibiotic activity of this organism in the late eighteen-eighties and the eighteen-nineties. The best known of these is the work of Emmerich and Low published in 1899, when they reported on the use of cell-free extracts from cultures of *Pseudomonas* for the topical treatment of infected wounds and claimed good results. They believed that the active substance was an enzyme, pyocyanase, though it was not clear what they thought was the substrate upon which this enzyme acted.

During the last decade of the nineteenth century a number of papers on the antibacterial action of fungi were published. Gasperini reported that *Strepthothrix foresteri* produced a substance that lysed cultures of various bacteria. In 1896 Gosio reported the preparation of crystals of a substance with antibiotic activity which he had isolated from the culture filtrates of a strain of *Penicillium*. This substance was later shown to be mycophenolic acid, a compound which is at present attracting some interest as a stimulator of interferon production in mammalian cells.

Thus by the turn of the century the existence of a number of microbial antagonisms had been demonstrated, and it had been shown in some cases that the effects were caused by diffusible antibiotic substances. Various strains of *Penicillium* and *Pseudomonas pyocyanea* had been the subject of particular attention, and the therapeutic possibilities were appreciated. Indeed Emmerich and Low's pyocyanase was produced commercially in Germany for a short time.

For the next twenty years little work seems to have been done on antibiotic substances of natural origin. This was probably due to the high hopes raised by the development of synthetic chemotherapeutic agents by Ehrlich and his followers. Trypan red, salvarsan and the acriflavines seemed to point the way to the development of effective antimicrobial therapy. The observations of Baudremer in 1913 on the antituberculous effect of extracts from the mould *Aspergillus fumigatus*, the description of the antibacterial effects of various actinomycetes by Leiske in 1921, and the work of Pringsheim and of Much and Sartorious on the antibacterial substances produced by members of the genus *Bacillus*, none of them excited any general interest, and serve only to show that the flame of interest though low and guttering, still burned.

Fig. 6.1. Sir Alexander Fleming (1881–1955)
in his laboratory at St Mary's Hospital,
London. By courtesy of St Mary's Hospital
Medical School.

The discovery of penicillin by Fleming in 1928 is a classical example of the role of serendipity in medical research. Fleming (Fig. 6.1) was studying the colonial morphology of staphylococci and by chance one of his plates became contaminated with a colony of the mould *Penicillium notatum*. Examining this plate under a lens he noted that the staphylococcal colonies around the large *Penicillium* colony showed signs of dissolution. Many bacteriologists might have thrown the contaminated plates away without further thought, but Fleming realized that here was an interesting example of microbial antagonism and proceeded to study the phenomenon further by intentionally seeding plate cultures of staphylo-

Fig. 6.2. Sir Howard Florey (1898–1968).
From J. Langdon-Davies, *Achievements in the
Art of Healing*, p. 19. London, The Pilot Press
(1945).

cocci and other organisms with point and streak inocula of
the *Penicillium.* The lytic effect upon the colonies of the
staphylococci and many other species of bacteria was confirmed.
Fleming then turned to the study of cell-free extracts of the
Penicillium cultures and found that these had a marked
antibacterial effect; he then proceeded to determine the toxicity
of the filtrates and found them to be harmless both to leucocytes *in vitro* and to experimental animals when administered by
injection in doses much larger than those needed to produce the
antibacterial effect. Fleming tried out the filtrates in the local
treatment of infected wounds with some success, but failing to
purify the material as a result of the lack of chemical facilities in
his laboratory he then abandoned the work for the time being.
The use of penicillin, as he called the antibacterial substance in

the filtrate of the mould cultures, did not become popular, and for the next ten years the only use of penicillin was in the preparation of a selective medium for the cultivation of *Bordetella pertussis*.

The next step forward was made in 1938–9 when Howard Florey, then Professor of Pathology at Oxford, began a systematic investigation of known antibacterial substances produced by microbes (Fig. 6.2). The penicillin of Fleming was one of the first to attract his attention. The known facts about its low toxicity and bactericidal effect obviously made it a worthwhile subject for study. Florey was fortunate in that he was backed up by a first-rate team of chemists headed by Ernest Chain. The fractionation of the crude filtrates of *Penicillium* cultures and the extraction and purification of the active substance proceeded apace. By 1940 preparations sufficiently pure for intramuscular injection in human subjects had been produced and the very low toxicity noted by Fleming was amply confirmed. It was then possible to proceed to the first clinical trials. Very little penicillin was available and only a few selected cases could be treated; however the results in cases of staphylococcal and streptococcal infections were so dramatic that there was never any doubt that penicillin was by far the most effective antibacterial drug yet found, and was capable of revolutionizing the treatment of many common infections. The situation of Britain in 1940 when a major war was in progress, at the same time made it urgent that large-scale production should be started so that the antibiotic be available for the treatment of the wounded and placed almost insuperable obstacles in the way of such a development. It was therefore decided to encourage the American pharmaceutical industry to undertake the further research and development. Florey and members of his team were flown to the United States and passed on the technical information that they had gathered over the past years and the American firms commenced work to develop techniques for the large-scale production of penicillin. There were two sides to the problem; in the first place mutant strains of *Penicillium* had to be isolated which were capable of growth in submerged culture and of producing an increased yield of penicillin, and in the second place there were the chemical engineering problems involved in the control of the environment in fermenters containing tens of thousands of gallons, and the separation and purification of the antibiotic from the fully grown cultures. The sterilization of the media required for such large vessels was a major problem. The American pharmaceutical industry was amazingly successful in solving these problems: high-yielding mutants capable of sub-merged growth were isolated and the engineering problems rapidly disposed of, so that within a year penicillin was available commercially and in ever-increasing quantities. At first it was only available for military casualties, but within months it became more and more widely available.

From the British point of view there was a heavy price to be paid for the rapid availability of the new drug. The American pharmaceutical firms that had done the research and develop-ment work naturally wished to protect their investment and patented the processes that they had developed. This has meant that British firms have, ever since then, had to pay royalties to these American firms in order to produce under licence an antibiotic that was a British discovery in the first place. This failure to exploit the industrial potentialities of a fundamental discovery is not unique: the story of the aniline dyes, their original synthesis by the British chemist Perkin and their development and commercial exploitation by the great German chemical firms in the latter half of the nineteenth century is an uncomfortably similar case, and one wonders whether we may not see another example in the case of the 'hovercraft'.

The discovery of penicillin led to a renewed interest in antibiotics from soil micro-organisms. In 1939 and 1940 Dubos had carried out some elegant work using the old principle of enrichment culture to isolate strains of bacilli from soil which exerted an antibacterial effect of Gram-positive cocci. The first substances discovered as a result of this work were gramicidin and tyrocidin, produced by Brewis (later work led in 1947 to the discovery of polymixin). Waksman (Fig. 6.3) and his group had been studying the microbial flora of the soil for years and in 1944 they announced the discovery of a new antibiotic, strep-tomycin, produced by *Streptomyces griseus*. This discovery came after a painstaking search which had involved the screening of no less than ten thousand different strains of bacteria, actinomy-cetes and fungi from soil samples. The new antibiotic seemed an ideal complement to penicillin, the former acting upon Gram-

Fig. 6.3. S. A. Waksman (1888–1973). From
S. A. Waksman, *My Life with the Microbes.*
London, R. Hale (1958). © Simon & Schus-
ter Inc., New York.

positive organisms, the latter upon the Gram-negative Enterobacteriaceae and the mycobacteria against which penicillin was
ineffective. In addition there was no cross-resistance between
the two compounds, so that in the case of those organisms
that were sensitive to both they could be substituted one for
the other in the event of a resistant strain emerging during
treatment. The most important result of the discovery of
streptomycin was the changed prognosis of tuberculosis. Instead
of years of rest and general supportive treatment in a
sanatorium aided in selected cases by surgical immobilization of
the affected organ, there was now a specific antibacterial therapy
which often produced a cure in a matter of months, and which
could be given if need be on an outpatient basis. Within a year or
so there were disturbing reports of patients who ceased to
respond to treatment and relapsed because their infecting

organisms had developed resistance to streptomycin, but even so the introduction of the drug had wrought a revolution in the treatment of tuberculosis. The solution of the problem of acquired resistance by the use of combined chemotherapy with streptomycin, *para*-aminosalicylic acid and isoniazid, followed a few years later and thus the promise of the early hopes was fulfilled.

Waksman's success led to a world-wide search for other antibiotic-producing soil micro-organisms sponsored by the great pharmaceutical companies. Soil samples from fields and compost heaps were collected from every part of the globe, pure cultures isolated from their microflora and tens, indeed hundreds, of thousands of isolates screened for antibiotic activity. Thousands of antibacterial compounds were purified and studied further, their spectrum of activity and toxicity being determined, and from all this work a few dozen therapeutic compounds were discovered. The few new antibiotics that did emerge were, however, a very important addition to the therapeutic armamentorium; they are often referred to as the broad-spectrum antibiotics. Chloramphenicol was first isolated in 1947, aureomycin in 1948, neomycin in 1949, terramycin in 1950 and tetracycline in 1953. These drugs not only acted upon a wide range of Gram-positive and Gram-negative bacteria, but upon the rickettsia and the infective agents of trachoma and lymphogranuloma inguinale.

The next important advance in our knowledge of antibiotics came in 1958 when Sheehan synthesized 6-aminopenicillanic acid, and thereby opened the way for the production of the semi-synthetic penicillins. Batchelor and his colleagues showed in 1959 that useful yields of penicillanic acid could be obtained from cultures of *Penicillium chrysogenum* grown in media to which no side chain precursor had been added. It was thus possible to obtain the nucleus of the penicillin molecule and by chemical means add any side chain that was desired. This approach has led to the production of a whole family of semi-synthetic penicillins custom-made to achieve certain ends: phenoxymethyl penicillin that is acid-resistant and so can be administered by mouth; methicillin that is unaffected by staphylococcal penicillinase and can be used to treat infections with 'penicillin-resistant staphylococci', and ampicillin which is both acid-resistant and has a broad spectrum of antibacterial activity inhibiting both Gram-negative and Gram-positive organisms, to name only the most important of these compounds.

The search for new antibiotics goes on; each year we read of new compounds that are being introduced. The position may be likened to a race between the bacteria which are constantly producing mutants resistant to the currently available antibiotics, and the microbiologists who are assiduously searching for new compounds to which the bacteria have not yet developed resistance.

REFERENCES

1874 Roberts, W. Studies on biogenesis. *Phil. Trans. R. Soc.*, Ser. B, **164**, 457–77.

1876 Tyndall, J. The optical deportment of the atmosphere in relation to the phenomena of putrefaction and infection. *Phil. Trans. R. Soc.*, Ser. B, **166**, 27–74.

1877 Pasteur, L. & Joubert, J. F. Charbon et septicémie. *C. r. Séanc. Acad. Sci.*, **85**, 101–15.

1885 Babes, V. & Cornil, A. V. Concurrence vitale des bactéries; atténuation de leurs propriétés dans des milieux nutritifs modifiés par d'autres bacteries; tentatives de thérapeutique bactériologique. *J. conn. Méd. Pract., Paris*, 3ᵉ Sér., **7**, 321–3.

1887 Baader, A. & Garré, C. Ueber Antagonisten unter den Bakterien. *Korrespbl. schweizer Aerzte*, **17**, 385–92.

1896 Gosio, B. Ricerche balteriologiche e chimiche sulle alterazioni del mais; contributo all'etiologia della pellagra. *Riv. d'ig. e. san. Publ. Roma*, **7**, 825, 869.

1899 Emmerich, R. & Low, O. Bakteriolytische Enzyme als Ursache der erworbenen Immunität und die Heilung von Infektionskrankheiten durch dieselben. *Z. Hyg. InfecktKrankh.*, **31**, 1–65.

1921 Leiske, R. *Morphologie und Biologie der Strahlenpily*. Leipzig: G. Borntraeger.

1929 Fleming, A. On the antibacterial action of cultures of a penicillium, with special reference to their use in the isolation of *B. influenzae. Br. J. exp. Path.*, **10**, 226–36.

1939 Dubos, R. J. & Cattaneo, C. Studies on a bactericidal agent extracted from a soil bacillus. *J. exp. Med.*, **70**, 1–10.

1940 Chain, E., Florey, H. W., Gardner, A. B., Heatley, N. G., Jennings, M. A., Orr-Ewing, J. & Sanders, A. G. Penicillin as a chemotherapeutic agent. *Lancet*, **ii**, 226–8.

1941 Hotchkiss, R. D. & Dubos, R. J. The isolation of bactericidal substances from cultures of *Bacillus brevis. J. biol. Chem.*, **41**, 155–62.

1944 Schatz, A., Bugie, P. & Waksman, S. A. Streptomycin, a substance exhibiting antibiotic activity against Gram-positive and Gram-negative bacteria. *Proc. Soc. exp. Biol. Med.*, **55**, 66–9.

1945 Florey, H. W. The use of micro-organisms for therapeutic purposes (a historical review up to 1945). *Br. med. J.*, **ii**, 635–42.

1949 Florey, H. W., Chain, E., Heatley, N. G., Jennings, M. A., Sanders, A. G., Abraham, E. P. & Florey, M. E. *Antibiotics.* London: Oxford University Press.

1958 Sheehan, N. C. *Aminoacids and Peptides with Antimetabolic Activity. Ciba Foundation Symposium*, p. 257. London: Churchill.

1959 Batchelor, F. R., Doyle, F. P., Naylor, J. H. C. & Robinson, J. N. Synthesis of penicillin: 6-aminopenicillanic acid in penicillin fermentations. *Nature, Lond.*, **183**, 257–8.

7

Microbial metabolism

Our knowledge of microbial metabolism has grown out of studies on the technical problems of industrial fermentations. For thousands of years the alcoholic and lactic fermentations have been in use for the production of wine, beer and cheese, whilst vinegar has been produced by aerobic decomposition of ethyl alcohol. Up to the middle of the nineteenth century there was no understanding of the nature of these processes; their control was an entirely empirical affair and when from time to time the fermentations went wrong and produced products that were at best unsaleable and at worst positively offensive, nothing could be done but scrap the batch, start again and hope that this time all would go well.

The first scientific paper concerned with microbial metabolism, Louis Pasteur's 'Mémoire sur la fermentation appelée lactique' of 1857, arose as a direct result of an investigation into the cause of repeated failures of fermentations intended to produce ethyl alcohol for industrial use. The facts of the case are interesting for the light that they shed on the relations between applied industrial research (what we should nowadays call 'trouble shooting') and fundamental advances in science.

In the summer of 1856 a certain manufacturer of alcohol in the town of Lille in northern France, a M. Bigo, suffered from repeated failures in his fermentations. The process involved the ferementation of beet sugar to produce ethyl alcohol, but time and again the contents of the vats turned acid and at the end of the process there was no alcohol, only a substance which smelled like sour milk. It happened that the son of M. Bigo was a student at the local Faculté des Sciences, a municipal college comparable to a polytechnic in England. The Dean of this institution was Louis Pasteur (Fig. 7.1), a young professor of chemistry who was

Fig. 7.1. Louis Pasteur (1822–1895). From an original photograph by Nadar, Paris. By courtesy of 'The Wellcome Trustees'.

known to be interested in industrial processes. M. Bigo approached Pasteur and asked him if he would undertake an investigation into the failures that were occurring in his fermentations. Pasteur agreed to do so and commenced his studies in his laboratory in the Faculté. In the first place he undertook a chemical analysis of the contents of the vats which had failed to produce alcohol. He was able to show that they contained a considerable quantity of lactic acid and no alcohol. It was fortunate that at this time there were methods available for the identification and estimation of lactic acid. The next step was to examine the sediment from the vats in which the fermentation had proceeded successfully and those in which it had failed. Pasteur had only a student's microscope available, but by this date even student's microscopes had achromatic objectives and were able to give good definition up to magnifications of about ×500. Comparison of the two sediments revealed a clear difference: in the sediments from the vats which had produced alcohol there were large yeast globules, some of which showed typical buds indicating that active growth was in progress; in the sediments from the vats which had produced only lactic acid there were 'globules much smaller than those of beer yeast' which were 'isolated or in groups of irregular masses' and 'moved actively by Brownian movement'. For some months Pasteur had been speculating upon the nature of fermentation. It is clear from his private notes that he was dissatisfied with the accepted account of the phenomenon in terms of a purely chemical instability. Now he had evidence that the products of these two fermentations were specifically associated with the growth of the two morphologically distinguishable microbes. Taking samples of the sediments from the two types of fermentations he inoculated them into tubes of sugar solution to which he added yeast extract in his laboratory. When he had added some chalk to the medium he was able to reproduce the lactic fermentation and to observe the minute globules in the grey sediment that collected above the chalk in the tubes. The addition of the sediment from the vats that had produced alcohol to a similar yeast extract/sugar solution, but without the chalk, resulted in a typical alcoholic fermentation and the appearance in the deposit at the bottom of the tubes of yeast globules. Thus was the foundation of the theory of fermentations as the metabolic activities of specific micro-organisms laid by Pasteur. The circumstances for this discovery were propitious: in the first place the products of the two fermentations were easily distinguished and methods for the chemical analysis and estimation of each of them were available; in the second place they were caused by two micro-organisms that could be easily differentiated on their microscopic morphology. This is not, however, any denigration of the brilliance of Pasteur in deducing the connection between the growth of the specific microbes and the production of the specific fermentation products, and further generalizing from these results a whole new theory of the nature of fermentation. In his paper to the Académie des Sciences announcing his discovery in 1857, Pasteur describes further experiments in which it was shown that the products of fermentation by the lactic acid organism, which he calls a yeast but which was almost certainly *Streptococcus lactis*, were altered by the addition or subtraction of chalk from the medium and thus for the first time demonstrated the effect of pH upon the metabolic activities of microbes. The 1857 'Mémoire sur la fermentation appelée lactique' of Louis Pasteur is rightly considered one of the classic papers on microbiology. It is not only that from it developed the whole edifice of microbial biochemistry and modern fermentation technology, but in the paper itself the seeds of a fantastic amount of future work can be discerned. For example, the specific nature of fermentations once established led to the concept of the specific causation of infectious diseases; the basic principles of media, the need for a carbon source, a nitrogen source, salts and vitamins are implicit in the recipe for the fermentation medium used; the alteration of the metabolic pattern of a micro-organism by the pH of the medium is a well established technique for driving a fermentation in a desired direction today, but it was a completely new idea at the time of this paper. Finally there is the suggestion that the acidity or alkalinity of the medium could be used to favour the growth of one or other species of microbe; the first hint of the use of selective media which now plays such an important part in the isolation of pathogenic bacteria from clinical specimens.

Not content with solving M. Bigo's problem and suggesting a general theory of fermentations, Pasteur went on to further studies, this time specifically directed to alcoholic fermentation. The results of this work were published in 1860 and gave a clear account of the role of yeast in the process, of the factors which influenced the fermentation and of the proportions of the different products.

As a part of further investigations into other fermentations Pasteur investigated the butanol fermentation. Here he found that it was not a yeast that was responsible, but a bacterium. Observations of drops of medium from a butanol fermentation revealed rod-shaped organisms that were actively motile in the centre of the drop but became immobilized at the edges of the drop where the oxygen tension was highest. This phenomenon interested Pasteur and led him to carry out further experiments which revealed that these rod-shaped bacteria were killed in the presence of oxygen, but moved actively and reproduced so long as they were kept under strictly anaerobic conditions. He reported his findings in a short paper to the Académie des

Sciences in 1861. While the phenomenon of microbes growing and moving in the absence of air had been observed by Spallanzani in the eighteenth century, Pasteur was the first biologist to appreciate the general significance of the observations.

The *mémoire* on alcoholic fermentation is historic because it was the first occasion when quantitative chemical data were used to solve a biological problem. There had been a long-standing controversy about the nature of fermentation and the relation of yeast to it. In 1835 and 1837 Caganard de Latour had published papers in which he had advanced the view that the yeast that sedimented out at the end of a vinous fermentation was 'a mass of globules that reproduced by budding and not merely a simple chemical or organic substance', and that it was 'very probable that the production of carbon dioxide and the decomposition of sugar and its conversion into alcohol are the effects of the growth of the yeast'. Unfortunately these views were opposed by Liebig and Berzelius, the two most eminent chemists of that generation, who while in disagreement as to the cause of fermentation were agreed in contemptuously rejecting the proposed biological explanation. Liebig believed that the fermentation of the sugar was due to the putrefaction of albuminous matter such as yeast communicating to other substances the state of decomposition in which it finds itself. Pasteur's accurate quantitative analyses showed that, in his own words: 'The nitrogen of the yeast is never transformed into ammonia during the alcoholic fermentation. Instead of ammonia formation, a slight amount of it disappears.' He was further able to demonstrate that yeast could be grown in a synthetic medium containing only sucrose, ammonium tartrate and yeast ash. During growth alcoholic fermentation took place and at the end of the process the ammonium salt had been used up, so that only minute traces could be detected in the medium. He went on to show that the yeast would not grow in the absence of either a fermentable sugar or a nitrogen source (ammonium salts or albuminous material). He concluded that 'fermentation can be accomplished under two separate circumstances, depending on whether the yeast is added to a solution of pure sugar, or whether the sugar solution is mixed with albuminous material. In the first case, the ferment acts but it does not reproduce. In the second case, it acts but it does reproduce.' He further concluded that 'Sugar never undergoes alcoholic fermentation without the presence of living globules of yeast.'

This paper conclusively proved the causal relation between the yeast and the fermentation of sugar to alcohol and finally disposed of the theories of Berzelius and Liebig. It was only possible to obtain such a decisive proof by the use of quantitative data and by developing arguments based thereon.

Later in the same year Pasteur published a paper in which he analysed the effect of oxygen upon the growth of yeast and the alcoholic fermentation. Once more quantitative methods were used to throw new light upon the problem. He was able to show that under aerobic conditions while the yeast reproduced vigorously its fermentative character was less marked, only four to ten parts of sugar being used up for each part of yeast produced. On the other hand, in the absence of oxygen only a small amount of yeast was formed, but a much larger amount of sugar was used up, sixty to eighty parts (by weight) disappearing for one part of yeast formed. In the discussion he speculates upon this phenomenon and states that it may be more generally found in other lower plants, thus making the first reference to the class of facultative anaerobes.

Pasteur now devoted himself to a systematic investigation of the technical problems involved in wine making. Wine making has always been an important industry in France and at that time was going through a very difficult time as a result of the death of many of the vines as a result of *Phylloxera* infestation. He visited vineyards and wineries all over France and noted the current practices in both fermentation and preservation. He was able to interpret these procedures in the light of the new theory of fermentations that he had developed, and to make many useful suggestions as to how the traditional techniques might be improved. His researches were published in a book entitled *Etudes sur le Vin, ses Maladies, Causes qui les provoquent; Procédés nouveaux pour le Conserver et pour le Vieillir* which came out in 1866. Amongst the improvements that he suggested was a method for improving the keeping quality of wines by heating them to a temperature of 68 °C for ten minutes and then rapidly cooling them. This technique has become famous as 'Pasteurization' and is now widely used for the treatment of milk to render it free from viable tubercle bacilli, brucellae and other pathogenic organisms.

The book on the diseases of wine was followed in 1871 by a book on 'the diseases of beer' in which the same procedure of analysis of current practices and suggestions for their improvement in the light of the new knowledge of the nature of fermentation was followed. It is perhaps significant that Pasteur used the terms 'diseases of wine' and 'diseases of beer' to describe the failures liable to occur in these fermentations. It suggests that even at this time he was thinking in terms of specific microbial causes of infectious diseases of men and animals.

In 1872 Ferdinand Cohn postulated the important role played by microbes in the biological cycles of the elements. Cohn, a botanist, appreciated the fact that there must be some mechanism by which the complex molecules synthesized by green plants from elementary materials were returned to the soil. The observed facts of the exhaustion of fields under continuous cultivation and the restoration of their fertility after the application of organic manures were the starting point for his argument and the new knowledge of chemistry and the microbial role in fermentations and putrefaction provided materials for the development of his ideas. Cohn had, of course, no suggestions to make as to the types of microbes which might be involved in these processes, but he was able to deduce their existence and to indicate that various micro-organisms might well play an important part in them. He was thus the first person to suggest a useful biological role for bacteria and fungi, and his theory stimulated later workers to search for and find not only the organisms responsible for the breakdown of complex organic molecules, but also the nitrogen-fixing and nutrifying bacteria.

The first fruits of Cohn's theory came in 1888 when the Dutch bacteriologist Beijerinck of Delft described the root nodule bacteria, *Rhizobium,* a symbiotic genus capable of fixing nitrogen when growing in the root system of leguminous plants. It had been known for some time that the roots of clover and related crops could be ploughed into the soil at the end of the season and would provide a very effective green manure. Indeed, such leguminous crops had been a regular part of the rotation of crops practised by progressive farmers since the middle of the eighteenth century. It was also known that these leguminous plants developed little nodules upon their roots, but what these nodules were was not known. Beijerinck's contribution was to show that the root nodules were little tumours caused by a localized bacterial infection of the root system. The bacteria were symbiotic, not only doing no harm to the plants but enabling them to grow in the absence of any nitrogenous salts. Seeds that were reared under sterile conditions failed to develop nodules and such plants were unable to grow in the absence of ammonium or other nitrogenous salts. It proved possible to infect plants grown under sterile conditions with bacteria from the root nodules of plants infected with the rhizobia and these then developed nodules on their roots and became able to grow with only elementary nitrogen available to them. The bacteria were able to fix atmospheric nitrogen, and so manufacture ammonium salts for the plant partner in the association, only when growing in the root nodules. In artificial culture they lost the ability to fix nitrogen. The relationship was thus a true symbiosis, both partners gaining from the presence of the other. The rhizobia are aerobes and it has been discovered more recently that they, alone amongst the bacteria, make use of haemoglobin in their respiratory pathway.

Beijerinck, as noted in a previous chapter, was also famous as the inventor of the 'enrichment culture' technique. It was a special application of this technique that led the Polish bacteriologst, Winogradski (Fig. 7.2), to the discovery in 1889 of a whole new class of bacteria, the autotrophs. These organisms are characterized by the ability to grow in media containing only carbon dioxide and inorganic salts. In some species this ability is facultative, but other species are obligate autotrophs and are unable to grow in the presence of complex organic molecules. Winogradski isolated the iron bacteria and the sulphur bacteria by inoculating soil samples into media which contained only iron salts or sulphates or sulphides and no organic carbon or nitrogen source. Under such conditions only autotrophic bacteria of the species that could utilize the inorganic constituents in the media could grow. Using such methods to isolate and grow pure cultures of these organisms Winogradski was able to recognize species that lived by the oxidation of ferrous to ferric salts, such as *Gallionella*, and a series of sulphur bacteria some of which, like

Fig. 7.2. Winogradski (1836–1953). From the slide collection of the Department of Bacteriology and Virology, University of Manchester. Source unknown.

Thiobacillus thioxidans, get their energy by oxidizing sulphur to sulphate, and others which use sulphates as hydrogen acceptors and gain their energy by oxidizing them to hydrogen sulphide. The descriptions of the autotrophic iron and sulphur bacteria were published in 1889. In the following two years Winogradski described two species of nitrogen bacteria, *Azotobacter* and *Nitrobacter*; the first is capable of fixing atmospheric nitrogen as ammonia, the second of transforming nitrite into the more highly oxidized nitrate.

Winogradski's work thus gave substance to the hypothetical cycles proposed by Ferdinand Cohn. The organisms responsible for the re-cycling of nitrogen, sulphur and carbon were now isolated and their metabolism understood at least in part. A further benefit of the work of Beijerinck and Winogradski was to demonstrate the extraordinary versatility of the patterns of bacterial metabolism. By the eighteen-nineties it was known that as well as the heterotrophs there were chemoautotrophs and photoautotrophs, using a variety of inorganic molecules as hydrogen donors or hydrogen acceptors, to use a terminology which was not to come into use for another thirty or forty years.

It is important to realize that at this time all bacterial metabolism was known only in terms of the overall conversions that took place. The fermentable substances that could be attacked by a given species and the end-products were recognized, but nothing was known of the intermediate metabolism.

The first step towards an understanding of the steps and pathways involved in the intermediate metabolism of micro-organisms was made when, in 1897 Büchner showed that the fermentation of sugar to alcohol and carbon dioxide could be affected by a cell-free extract from crushed yeast. Superficially this might appear to subvert the theory of fermentation so ably developed forty years before by Louis Pasteur. His experiments appeared to demonstrate that fermentation only took place in the presence of living yeasts or other micro-organisms and were in fact a manifestation of their metabolic activity. Now it was shown that living yeast cells were not necessary for the process. Were Liebig and Berzelius correct after all? In fact there was no such refutation of Pasteur's theories implicit in the observations. Büchner's discovery is a classical case of that dialectical progress likened to a spiral by the philosopher Hegel. Büchner's discovery was the antithesis of Pasteur's work on fermentation, but it occurred at the next level of historical development and therefore led to a synthesis at this higher level.

The idea that there might be special proteins within yeast cells responsible for the fermentation of sugar had been suggested as long ago as 1858 by Traube, but up to the date of Büchner's work all attempts to extract and isolate such 'enzymes' had been unsuccessful. Since these enzymes were only produced by living yeast cells there was no conflict with the theory of Pasteur: all that was being done was to define the particular part of the yeast cell which was responsible for the fermentative activity.

Büchner ground up a kilogram of brewer's yeast with quartz sand and kieselguhr and then placed it in a filter press and subjected it to a pressure of 400–500 atmospheres. From one kilogram of yeast he obtained 500 cubic centimetres of cell-free juice. The most interesting property of this juice was its ability to bring about the fermentation of carbohydrates. Büchner states 'when an equal volume of the juice is mixed with a concentrated solution of cane sugar, after only a quarter to one hour the regular formation of carbon dioxide bubbles begins, and this may last for many days...after three days fermentation at icebox temperature, the alcohol formed was determined. In one experiment with 50 cc. of yeast juice 1.5 gms of ethyl alcohol was formed.'

Büchner thought that a single enzyme was responsible for a one-step conversion of glucose to ethyl alcohol. He even named it zymase. Shortly afterwards in 1900 the work of the German biochemist Neuberg shed further light on the problem and showed that the fermentation of glucose to ethyl alcohol was accomplished not in a single step by a single enzyme, but in numerous small steps each requiring its proper enzyme. It was to take a number of years before the full complexity of the pathway and the various enzymes and co-enzymes were fully described, but at last the right model had been proposed and a true understanding of the stepwise metabolic pathways of living organisms was possible.

It was also in 1900 that Wieland first put forward the thesis that biological 'oxidations' were in fact accomplished by the removal of hydrogen atoms from some hydrogen donor and their transfer to one of a number of alternative hydrogen acceptors, of which oxygen was the commonest but by no means the only example. This view of the energy-producing mechanisms of living organisms was to prove very fruitful. Not only was it to lead to the identification of many dehydrogenating enzymes by Thunberg and his followers, but in the hands of the eminent Dutch microbiologist Kluyver it was to provide the key to a demonstration of the essential unity that underlies the apparent diversity of the metabolic patterns of microbes. It is seldom in biology that so productive an idea has been brought forward. The only parallels that come to mind are the theory of evolution of Darwin and the concept of the high-energy phosphate bond put forward in the early nineteen-forties by Lipmann and his colleagues.

During the next decade the metabolism of carbohydrates by

other micro-organisms was studied from the point of view of the intermediate metabolism and it became clear that in all cases the process proceeded by discrete steps each catalysed by a specific enzyme.

Another approach to the problem was provided by the work of Harden and his collaborators who studied the products of the metabolism of glucose under aerobic and anaerobic conditions by the closely related bacteria, *Escherichia coli* and *Aerobacter aerogenes*. Their work demonstrated the flexibility of bacteria, which are able to switch from one pathway to another according to the environmental conditions, and also shed light upon the nature of the differences that underlie the diagnostic tests used to differentiate between the free-living *A. aerogenes* found in soil and the commensal *E. coli* characteristically found in the gut of man and animals.

The accumulation of comparative biochemical data on various bacteria led Orla Jensen, the Danish agricultural bacteriologist, to put forward a new scheme for the classification of the schizomycetes based upon metabolic characters. All previous taxonomic schemes had been based upon morphological criteria, with a few biochemical characters put in to distinguish between species which were morphologically identical. Orla Jensen attempted to go behind the morphological phenomena and divide bacteria into families, genera and species based on their types of metabolism. The imperfect state of knowledge prevented his scheme being altogether satisfactory and much of it has been abandoned by modern workers in the field of taxonomy, but certain entities which he established, such as the family Lactobacteriaceae, have become permanent features of all later classifications, and whatever its limitations his scheme was a major advance in the approach to bacterial classification.

Up to 1917 the regulation of the reaction of bacteriological media had been a hit-and-miss affair. It had been realized from the first, as can be seen in Pasteur's 1857 paper on the lactic fermentation, that some bacteria grew best in acid media and some in alkaline media, and that a change in the reaction of the medium could be used to select a desired organism or to suppress an undesirable one. There was, however, no quantitative method of expressing the degree of acidity in use; it could only be said that the medium should be very acid, slightly acid,

Fig. 7.3. Marjorie Stephenson (1885–1948).
From *J. gen. Microbiol.* (1949), **3**, facing
p. 329.

neutral, slightly alkaline or very alkaline. In 1917 the Americans,
Clark and Lubs, published their classical paper in the *Journal of
Bacteriology* entitled 'pH, its measurement and significance in
bacteriology'. By applying the quantitative notation developed
some years before by physical chemists they were able to define
the optimum degree of acidity for the growth of species of
organisms with precision, and the colorimetric methods for
measuring pH that they introduced into bacteriology made it
possible for the first time to produce media with a standard
reproducible degree of acidity or alkalinity. Clark and Lubb's
paper produced a revolution in media making, and as a
secondary result methods for the more rapid isolation of a
number of pathogenic organisms were developed, as well as
fresh selective media. Perhaps an even more important result of
their paper was to make bacteriologists 'pH conscious' and

provide a stimulus to work on the pH optima of various bacterial enzymes in an attempt to understand the precise mechanisms that determined the optimal pH for the growth of given species of microbe.

In the early nineteen-twenties there was a very active group of biochemists working on bacterial metabolism in Cambridge. The most famous of these workers were Marjorie Stephenson (Fig. 7.3) and J. H. Quastel. Their contribution was to the study of the enzymes of bacteria and in particular to the study of the dehydrogenases. Applying the technique of Thunberg (which involves the measurement of the time taken for the decolorization of methylene blue in evacuated tubes in which are placed a hydrogen donor, a suspension of bacteria and the dye which acts as a hydrogen acceptor), they made a systematic study of the dehydrogenases of a number of bacteria. In the first place they were able to identify dehydrogenases specific for a number of fatty acids, alcohols and amino acids, and in the second place they were able to determine the pH optima and Michaelis constants of these enzymes. From such information it was possible to build up a picture of the various pathways of intermediate carbohydrate metabolism utilized by different species of bacteria and a beginning was made in the understanding of the way in which bacteria could switch from one pathway to another according to the environmental conditions, pH, oxygen tension, and temperature, or as a response to different substrates. It might be said that the work of Stephenson and her associates established the enzymatic anatomy of the catabolic pathways in bacteria. Little or nothing was known at this time of the synthetic pathways in bacteria, and the problem of how the energy produced by the catabolic pathways was stored and converted into work was still unsolved.

In 1924 Kluyver (Fig. 7.4) published an important article entitled 'Unity and diversity in the metabolism of micro-organisms'. Drawing upon the results that were by now available for a large number of bacterial species, he showed how the superficial diversity of energy-producing mechanisms, aerobic and anaerobic, heterotrophic and autotrophic, were the result of different combinations of similar enzymes and co-enzymes, so that there was a fundamental unity which underlay the apparent diversity. Kluyver had succeeded Beijerinck as the head of the

Fig. 7.4. A. J. Kluyver (1888–1956). From
Biogr. Mem. Fellows R. Soc. (1957), **3**, facing
p. 109.

School of Microbiology at Delft and was developing a wide-
ranging programme of research in the field of comparative
biochemistry, a term which he himself coined some years later
in his London lectures.

In 1930 Karstrom first distinguished between constitutive and
adaptive enzymes in bacteria. Up to this time it had been
believed that in bacteria, as in other types of cells, enzymes were
either present or absent, and that the only way that an enzyme
could be gained or lost was by genetic mechanisms. Karstrom
showed that in bacteria, in addition to the constitutive enzymes,
there were others which were not detectable, or present in very
low concentration in the cells, until their substrate was added to
the medium, when they appeared in increasing amounts over a
short period of time and in the absence of cell multiplication.
These adaptive enzymes could only be produced by synthesis
within the cell, and such synthesis appeared to be switched on by
the presence of the appropriate substrate, acting as an inducer.

It was obvious from the start that these inducible or adaptive enzymes were very important mechanisms for bacteria, with a high survival value. The ability to adapt to an altered environment without the cumbersome mechanisms of mutation and selection but within the lifetime of the affected cells, is one that is found only in the animal kingdom apart from bacteria.

The discovery of the antibacterial action of the sulphonamides in 1935 stimulated interest in their mode of action. In 1940 D. D. Woods and Sir Paul Fildes elucidated the problem and in so doing uncovered a mechanism of very great importance for our understanding not only of drug action but of enzyme kinetics in general. Their hypothesis was that the sulphonamide molecule acted by competitive inhibition of an enzyme whose normal substrate was *para*-amino benzoic acid (PABA), a molecule with close structural affinities with sulphonamide. Their experiments showed that the antibacterial action of sulphonamide could be reversed by the additon of PABA to the medium, and more important that the reversal was stoichiometric, fifty parts of sulphonamide being neutralized by one part of PABA. This discovery raised hopes that a fruitful approach to bacterial chemotherapy might be the synthesis of antimetabolites. Many such compounds were investigated but little progress was made in the years that followed. Some of the anti-tumour drugs used in the treatment of leukaemia are the product of this line of research, but no effective antibacterial drugs have been discovered in this way. It is interesting to speculate as to why this should be so. The approach is rational, and the first antimetabolite, sulphanilamide, has proved to be an extremely effective antibacterial agent. Why then did the later antimetabolites prove ineffective? It is only lately that we have come to understand the way in which the metabolic activities of the bacterial cell are regulated at the molecular level: the negative feedback mechanisms of allosteric inhibition and repression and the interactions between different pathways were not appreciated thirty years ago; had they been there would have been less hope that the addition of a single antimetabolite competing for a single enzyme would be likely to produce an antibacterial effect.

In 1941 a very important paper by Lipmann and his colleagues was published. In it they advanced the concept of the high-energy phosphate bond. It had been known for many years

Fig. 7.5. Professor G. W. Beadle (1903–   ).

that phosphate was fixed during the process of fermentation, and it was later shown that inorganic phosphate concentration in the medium was a rate-limiting factor in bacterial and fungal fermentation and respiration. It had also been known that inorganic phosphate was liberated by contracting muscle in the state of 'oxygen debt' and consumed by the same tissue during the recovery phase. The organic compounds that were formed by cells – adenosine mono-, di-, and triphosphate – were well known and it was believed that they acted as energy stores. Lipmann's group brought all this information into focus, by postulating that certain phosphate–phosphate bonds had the property of releasing much more energy on hydrolysis than simple phosphate bonds. The classic example of such high-energy phosphate bonds are the terminal and subterminal bonds between the phosphate groups in adenosine triphosphate (ATP).

The studies of Knox and Pollock in England and Monod in France during the late nineteen-forties clarified the picture that we have of the mechanisms involved in the induction of adaptive enzymes. The work of Monod in particular was most elegant and

established the new fact that substances which were not substrates for the enzyme could act as inducers. Indeed he found
that in the case of β-galactosidase in *E. coli* the most powerful
inducer was not a substrate for the enzyme.

In about 1945, studies on the synthetic pathways of bacteria
began to appear in the literature in large numbers. Up to this
time much more had been known about catabolic processes in
bacteria than about anabolic processes. The new work was
stimulated by the application of techniques for the isolation of
nutritionally defective mutants that had been worked out over
the last few years by Beadle and his school at Stanford
University, California, using the red bread mould *Neurospora*.
Beadle (Fig. 7.5) and his associates were geneticists and the
techniques were designed with such studies in mind, but the
methods of biochemical genetics were equally applicable to
the study of the synthetic pathways in their own right. Once the
nutritionally defective mutants produced by ultra-violet or X
irradiation had been isolated, the precise nature of the block and
its place in the synthetic pathway could be determined by the use
of chemical analysis of the products accumulated by such strains,
and cross-feeding experiments. Ultimately a series of mutants
each defective in one step of the synthetic pathway could be built
up. In this way, the complete synthetic pathways of amino acids
such as tryptophan, and of more complex substances such as
folic acid were elucidated, so that by the end of the decade almost
as much was known about the synthetic as about the energy-
producing mechanisms of bacteria.

The use of radioactive isotopes was another powerful technique for the investigation of the intermediate metabolism of
bacteria. The feeding of glucose labelled with ^{14}C at a specific
atom, followed by the examination of the products of fermentation for the presence of labelled carbon atoms, led to the
recognition of new glycolytic pathways. The converse approach,
the feeding of precursors labelled with isotopic atoms, led to a
fuller understanding of many synthetic pathways. In the case of
the synthetic pathways the radioactive tracer studies were
complementary to the studies using defective mutants and rapid
progress was made during the late nineteen-forties and early
nineteen-fifties.

The year 1961 saw the birth of molecular biology. The triplet code relating nucleotides to amino acids was published by Crick and his associates working in Cambridge. Later in the year Niremberg and Matthaei announced the successful programming of *E. coli* ribosomes for the production of polyphenylalanine by the use of the synthetic polynucleotide polyuridine. Thus the basis for our further understanding both of the replication of DNA and the role of RNA in protein synthesis and the understanding of the genetic code was laid.

Since then these foundations have been built upon with great rapidity. The genetic code is now almost, if not completely, understood, the details of the mechanism and control of protein synthesis have been unravelled in procaryotic cells, and molecular biologists are beginning to attack the more complex problems of the control of protein synthesis in eucaryotic cells and the mechanisms responsible for the differentiation of tissues in multicellular organisms.

REFERENCES

1857 Pasteur, L. Mémoire sur la fermentation appelée lactique. *Mém. Soc. Sci. Agric. Arts*, Lille, **5**, 13–26.

1857 Pasteur, L. Mémoire sur la fermentation appelée lactique. *C. r. Acad. Sci.*, **45**, 913–16.

1857 Pasteur, L. Mémoire sur la fermentation alcoolique. *C. r. Acad. Sci.*, **45**, 1032–6.

1858 Pasteur, L. Mémoire sur la fermentation appelée lactique. *Ann. Chim. Phys.*, 3ᵉ sér., **52**, 404–18.

1860 Pasteur, L. Mémoire sur la fermentation alcoolique. *Ann. Chim. Phys.*, 3ᵉ sér., **58**, 323–426.

1861 Pasteur, L. Animalcules infusiores vivant sans gaz oxygene libre et determinant des fermentations. *C. r. Acad. Sci.*, **52**, 344–7.

1866 Pasteur, L. *Etudes sur le vin, ses Maladies, Causes qui les Provoquent, Procédés Nouveaux pour le Consérver et pour le Vieillir.* Paris (1866).

1871 Pasteur, L. *Etudes sur la Bière, ses Maladies, Causes qui les Provoquent, Procédé pour la Rendre inalterable avec une Théorie Nouvelle de la Fermentation.* Paris (1876).

1872 Cohn, F. *Bacteria, the smallest Living Organisms.* Translated by C. S. Dolley. Johns Hopkins Press (1881).

1888 Beijerinck, M. S. Die Bakterien der Papilionaceen-Knöllchen. *Bot. Ztg.*, **46**, 724–35.

1889 Winogradski, S. Recherches physiologiques sur les sulphobacteries. *Ann. Inst. Pasteur*, **3**, 49–60.

1890 Winogradski, S. Recherches sur les organismes de la nitrification. *Ann. Inst. Pasteur*, **4**, 213–31, 257–75, 760–71.

1891 Winogradski, S. Recherches sur les organismes de la nitrification. *Ann. Inst. Pasteur*, **5**, 92–100, 577–616.

1897 Buchner, E. Alkoholische Gährung ohne Hefezellen. *Ber. dtsch. chem. Ges.*, **30**, 117–24.

1904 Harden, A. *Alcoholic Fermentation.* London: Longmans, Green & Co. (Fourth edition 1932.)

1907 Stockhausen, S. Okologie, 'Anhau-fungen' nach Beijerinck. Berlin: Institut für Garungsgewerbe.

1909 Jensen, O. Die Hauptlinien des naturlichen Bacteriensystems. *Zentralbl. Bakt.*, **22**, 305–46.

1913 Wieland, H. Uber den Mechanismus der Oxydationsvorgänge. *Ber. dtsch. chem. Ges.*, **46**, 3327–42.

1917 Clark, W. M. & Lubs, H. A. The colorimetric determination of hydrogen ion concentration and its applications in bacteriology. *J. Bact.*, **2**, 1–34.

1924 Kluyver, A. J. Eenheid en Verscheidenheid in der Stofwisseling der Microben. *Chem. Weekbl.*, **21**, 266–77.

1924 Quastel, J. H. & Whetham, M. D. The equilibria existing between succinic, fumaric and malic acids in the presence of resting bacteria. *Biochem. J.*, **18**, 519–34.

1925 Quastel, J. H. & Whetham, M. D. Dehydrogenations produced by resting bacteria: I. *Biochem. J.*, **19**, 520–31.

1925 Quastel, J. H. & Whetham, M. D. Dehydrogenations produced by resting bacteria: II. *Biochem. J.*, **19**, 645–51.

1929 Fleming, A. On the antibacterial action of cultures of a penicillium, with special reference to their use in the isolation of *B. influenzae. Br. J. exp. Path.*, **10**, 226–36.

1930 Karstrom. Ueber die Enzymbildung in Bakterien. Thesis, Helsingfors.

1931 Kluyver, A. J. *The Chemical Activities of Micro-organisms.* London University sities Press.

1935 Domagk, G. Ein Beitrage zur Chemotherapie der Bakteriellen Infektionen. *Dtsch. med. Wochenschr.*, **61**, 250.

1940 Woods, D. D. The relation of *p*-aminobenzoic acid to the mechanism of the action of sulphanilamide. *Br. J. exp. Path.*, **21**, 74–90.

1944 Knox, R. & Pollock, M. R. Bacterial tetrathionase: adaption without demonstrable cell growth. *Biochem. J.*, **38**, 299–304.

1947 Monod, J. The phenomenon of enzymatic adaptation and its bearings on problems of genetics and cellular differentiation. *Growth*, **11**, 223–89.

1948 Pollock, M. R. Penicillinase adaptation in *B. cereus*; adaptive enzyme formation in the absence of free substrate. *Br. J. exp. Path.*, **31**, 739–53.

1948 Pollock, M. R. & Wainwright, S. D. The relationship between nitratase and tetrathionase adaptation and cell growth. *Br. J. exp. Path.*, **29**, 223–40.

1961 Crick, F. H. C. *et al.* General nature of the genetic code for proteins. *Nature, Lond.*, **192**, 1227–32.

1961 Nuremberg, M. W. & Matthaei, J. H. The dependence of cell-free protein synthesis in *E. coli* upon naturally occurring or synthetic polyribonucleotides. *Proc. nat. Acad. Sci. USA*, **47**, 1588–1602.

8

Microbial genetics

The study of bacterial genetics derives from earlier studies on 'bacterial variation' and this again can be seen to be continuous with the controversies concerning spontaneous generation. Proof of the stability of bacterial species would have been one of the most powerful pieces of evidence that could have been produced against the hypothesis of heterogenia or spontaneous generation. Such evidence, however, was almost impossible to obtain before methods for the production of pure cultures were available. Spallanzani, working only with a microscope, described the different types of animalcules that he saw in the infusions which underwent putrefaction, but the multiplicity of types which he recorded were used by his opponents to support their theory that they all arose from the action of some vital force on inanimate matter. Louis Pasteur in his famous controversy with Pouchet between 1860 and 1865 was more fortunate in that he had already demonstrated the specific relation between certain microbes and definite fermentations. He was still, however, bedevelled by the difficulties of obtaining pure cultures using only liquid media. The problem though difficult was not insoluble and by 1870 Pasteur and Ferdinand Cohn, working the one in France and the other in Germany, had convinced the scientific world that there were stable bacterial species: that cocci gave rise to cocci, bacilli to bacilli and vibrios to vibrios. This concept of stable species was further strengthened when in 1882 Robert Koch introduced the use of solid media, thus enabling pure cultures to be obtained by the simple method of streaking out a mixed culture on a plate and picking off single colonies.

During the last two decades of the nineteenth century evidence began to accumulate that bacterial species were not as stable as had at first been thought. Pure line cultures which had been maintained in the laboratory for many generations were reported to have suddenly undergone dramatic changes in colonial morphology, fermentative ability or pathogenicity.

As the methods of examination of bacterial cultures became more varied the number of cases of such variation, or dissociation as it was called, multiplied, but always these phenomena were regarded as abnormal occurrences which were to be guarded against, not as subjects worth investigation on their own account.

In 1907 the first study of bacterial variation for its own sake appeared; Massini's paper on *Bacterium coli mutabile* described the occurrence of lactose-fermenting colonies in the non-lactose-fermenting strain of the organism. For the first time a quantitative approach was used and the frequency of these variant colonies was estimated. It was shown that such lactose-fermenting colonies bred true, and it became clear that this type of bacterial variation was in many ways like the mutations observed in higher organisms. At that date it was not possible for it to be recognized as a mutation because the bacteria were regarded as having no genetic apparatus comparable to the nucleus in higher cells. Massini's paper remains a classic example of how to investigate bacterial variation at the phenotypic level.

Cases of variations continued to be reported and bacteriologists naturally speculated upon the cause of such variations. In 1925 R. M. Mellon published a paper 'Observations on a primitive form of sexuality in the colon–typhoid group'. The views put forward could not be tested experimentally at that time and his speculations had little influence upon contemporary thought. In the view of almost all bacteriologists the bacteria were anucleate organisms, reproducing by binary fission and without any sexual stage in their life cycle.

The year 1927 saw the publication of a monumental study of 'microbial dissociation' by Phillip Hadley in the *Journal of Infectious Diseases*. In this paper, which runs to some three hundred pages, Hadley reviews almost every type of variation or, as he calls it, dissociation, which had ever been reported as affecting bacterial cultures. Variations in colonial and microscopic morphology, fermentations, and virulence, variations in optimal temperature for growth and in resistance to various antiseptic substances, are all noted. It is an interesting exercise to read through Hadley's article and try to interpret his observations in terms of modern genetic concepts. Many of the types of variation that he reports are only explicable by the occurrence of contamination of the cultures, but a substantial proportion appear to have been cases of mutation or of the selection of phenotypes present in very small numbers in the populations studied.

The paper published by Griffith (Fig. 8.1) in 1928 describing the phenomenon of 'transformation' in cultures of pneumococci is important not only for its immediate effect, but even more so because it was the foundation upon which Avery and his colleagues were to later establish the role of deoxyribonucleic acid (DNA) as the molecule on which the genetic information of all cells is coded. The phenomenon that Griffith described, the heritable change of rough, avirulent type I pneumococci to smooth, virulent typle II pneumococci, effected by growing them in the presence of heat-killed smooth type II pneumococci, was in itself of great interest, but its implications for our understanding of the chemical basis of heredity were even greater.

A further step in the understanding of transformation was made in 1933 when Alloway showed that rough type I cells could be changed into genetically stable smooth type II cells by growing them in the presence of a cell-free filtrate prepared from a heat-killed broth culture of smooth type II cells, thus proving that the 'transforming agent' was a soluble, or at any rate a subcellular, material.

In the same year Reed published a paper in the *Journal of Bacteriology* postulating an unequal segregation of genetic material at the time of cell division as the mechanism of bacterial variation. As bacteria were at that time thought to be anucleate, their genetic material being diffused throughout the cytoplasm, this was not an unreasonable hypothesis.

During the early nineteen-forties a new interest in bacterial variation was awakened. The use of antibiotics in medicine led rapidly to the emergence of strains of bacteria resistant to these agents and the attendant clinical problems. Here was a type of bacterial variation that was of obvious clinical significance, and medical bacteriologists all over the world turned their attention

Fig. 8.1. Frederick Griffith (1879–1941).
From *J. gen. Microbiol.* (1966), **45** facing
p. 385.

to the elucidation of its mechanisms. This occurred just at the time when two major advances had been made in the field of genetics. In the first place Beadle and his associates at Stanford University in California had, since 1941, been developing techniques for studying the genetics of the red bread mould *Neurospora crassa* using biochemical characters as phenotypic markers. Beadle had for some years been working in the field of genetics studying the fruit fly *Drosophila*. He turned to fungal genetics because it promised two advantages: a shorter generation time, so that crossing experiments could be completed in days rather than weeks, and the possibility of examining the enzymatic changes that were known to underlie the morphological phenotypic characters recorded in great detail in *Drosophila* and other higher animals and plants. These techniques for the study of biochemical genetics in fungi were immediately applicable to bacterial cultures. The first fruits of their application came in 1943 when Luria and Delbruck showed by means of

Fig. 8.2. O. T. Avery (1877–1953).

their 'fluctuation test' that spontaneous mutations occurred in bacteria, to both phage resistance and streptomycin resistance, at the same sort of frequencies as had been observed in other organisms. In the second place in 1944 Avery (Fig. 8.2) and his colleagues re-examined the phenomenon of transformation in pneumococci, using modern chemical techniques to identify the nature of the 'transforming principle'. By a series of elegant experiments they showed that the molecule responsible for the transformation was DNA. This was a revolutionary discovery. Up to that time geneticists had believed that the significant part of the nucleo-protein of the chromosomes was the protein, the nucleic acid acting as a sort of glue to stick the essential protein material together in the correct way. The identification of DNA as the compound in which the genetic information was coded was as puzzling as it was unexpected, for it was difficult to see

Fig. 8.3. Professor Joshua Lederberg
(1925—). By courtesy of the News and
Publications Service, Stanford University,
California.

how a polymer that contained only four bases could possibly
code for the very large numbers of characters that go to make up
the phenotype of even the simplest of organisms. It was not until
some fifteen years later that the problem was solved and the
general nature of the code was discovered. However, interest
was from now onwards directed to DNA and over the next few
years an enormous amount of information about its physical and
chemical structure was accumulated. It was this information
which made it possible to solve the code in due course.

Meanwhile the classical genetic approach was leading to new
discoveries. In 1945 Tatum showed that, as with other organisms, the mutation rate of bacteria could be increased by treating
them with X-rays. In 1947 Tatum and Lederberg (Fig. 8.3)
demonstrated genetic recombination between two nutritionally
defective strains of *Escherichia coli* K12. They were able to show
that after incubating the two parent strains together in a full

liquid medium, cells appeared which could grow upon a deficient medium which failed to support the growth of either of the parent strains. It was further shown that these cells had characters which could only have arisen by their incorporating genetic material from both parents. This phenomenon was further studied by Davis who showed in a very elegant experiment that cell to cell contact was necessary for it to occur, thus ruling out transformation as an explanation.

By 1946 it was becoming possible to explain most of the phenomena of bacterial variation, and in particular the emergence of strains resistant to antibiotics, in terms of mutation and selection. There were, however, a number of influential micro-biologists who were unwilling to accept such a hypothesis, preferring to explain the observed facts in terms of a mysterious process referred to as 'training'. The most able exponent of the 'training' theory was Sir Cyril Hinshelwood, an Oxford Professor of Chemistry, and a man of encyclopaedic learning. The breadth of his knowledge was such that at a later date he had the unique privilege of being both the President of the Royal Society and the President of the Classical Association. Hinshelwood's book *The chemical kinetics of the bacterial cell*, a neo-Lamarckian work explaining the emergence of resistance to antibacterial substances and other types of variation in terms of adaptation by individual cells, was for a time very influential. While the theory upon which it was based was incorrect, its ultimate effect was beneficial as it forced the supporters of the mutation and selection hypothesis to marshal their evidence to such effect that their case was finally conceded by their most ardent opponents.

The year 1947 saw the publication of the first gene map for a bacterium, *E. coli* K12. While not complete, the loci for several dozen genes were defined by the use of linkage analysis following crosses.

During the next few years bacterial genetics advanced in two directions. In the first place conjugation, transduction and transformation were shown to occur in a number of different species of bacteria and not to be peculiar to the strains in which they had been first described. In the second place, new facts concerning the mechanism of conjugation were discovered. In 1947 it was shown that antigenic variation could be produced by means of transformation in both *E. coli* and in *Shigella* species. In

1950 it was reported that it was possible to confer motility upon *Bacillus anthracis* by transformation with DNA from other species of the genus *Bacillus*. In 1951 it was announced that it had been found possible to produce antigenic changes in *Haemophilus influenzae* by transformation. In 1952 transduction with a phage affecting *Salmonella typhimurium* was described. In 1953 recombination was described for the first time in a virus, a bacteriophage. In 1955 conjugation and recombination were reported in *Pseudomonas pyocyanea*, and in 1956 in *Serratia marcesans*.

A most important contribution to the understanding of the mechanisms of recombination following conjugation was made by William Hayes, then working at the postgraduate Medical School in Hammersmith. In a paper in *Nature* in 1952 he announced his discovery that in conjugation, recombination took place as a result of a one-way transfer of genetic material from donor cells. In the same year the Lederbergs and Cavalli confirmed this and coined the terms fertility plus (F+) for donor cells and fertility minus (F−) for receptor cells. Up to this time it had been uncertain what happened during the process of recombination, and it seemed possible that a two-way exchange of genetic material might take place between cells of a similar mating type. The recognition of the F+ and F− mating types made it clear that conjugation was a primitive type of sexuality with the recipient cell (F−) being the zygote. This discovery opened the way for rapid advance in our knowledge of the bacterial chromosome during the following decade.

In parallel with these discoveries progress was made in other fields related to bacterial genetics. In 1949 Robinow demonstrated that bacteria contained chromatinic bodies which stained positively with Feulgen's reagent provided that the cells were first treated with warm hydrochloric acid to hydrolyse all the RNA present. He deduced, quite correctly, that these masses of DNA were the analogues of the nucleus in higher cells. Thus the nuclear apparatus which had been hypothesized on purely genetic data was for the first time visualized under the microscope. In 1952 there was the publication by Lederberg and Lederberg of an extremely powerful and simple method for detecting mutants in bacterial populations, the technique of replica plating. By means of this technique, it was possible to demonstrate unequivocally that mutants resistant to drugs occurred with a definite frequency in the absence of the drug.

The discovery of high frequency recombinant mutants arising from the F+ type of *E. coli* K12 was first reported in 1952 by Lederberg, Cavalli and Lederberg and confirmed by William Hayes in 1953. These mutants strains which they called Hfr strains, differed from the wild-type F+ strains not only in transferring various genetic markers at a rate hundreds of times greater than the original strains, but also in not producing an alteration in the mating type of the recipient cells. The frequency of transfer of the various markers differed but was the same for any given Hfr strain.

Between 1955 and 1958 Jacob and Wollman elucidated the mechanism of gene transfer in *E. coli* K12 in detail in a series of ingenious studies using Hfr strains. The most famous and critical of their studies was the 'interrupted mating experiment'. In this type of experiment an Hfr and and an F− strain, both in the early logarithmic phase of growth, are mixed in a well aerated medium containing glucose as an energy source. At intervals after the moment of mixture samples are withdrawn and treated in a high-speed blender before being plated out on solid media suitable for the selection of recombinants. This treatment severs the conjugation tube between pairs of organisms thus terminating the exchange of genetic material, but does not kill the cells. Experiments of this type showed that the act of conjugation was complete in about one hour, and that the number of markers transferred was greater the longer the pairs were left conjoined before severance in the blender and subsequent culture. Thus it was clear that the transfer of genetic information was not only unidirectional but linearly orientated. Gene maps constructed from this 'time analysis' were in conformity with those constructed using classical genetic data on crossover frequencies.

Further evidence about the detailed mechanism of conjugation and zygote formation was provided by the use of the replica plating technique with which it was possible to isolate numerous different Hfr mutants from the same F+ culture. It was found that these various mutants transferred their markers each in a different order. However the order for a given Hfr mutant was constant. A comparison of the different orders led to the recognition that the only model that would fit the facts was that of a circular chromosome which was opened at different points by the insertion of the F factor in the various mutants. Such a model postulated an F factor which could exist independently of the chromosome in the F+ state of the cell and alternatively integrated into the chromosome in the Hfr state, and further which was transferred in isolation following conjugation between F+ and F− cells, but was the last part of the chromosome to be transferred in conjugation between Hfr and F− cells.

Jacob and Wollman in 1958 coined the term 'episome' to describe this class of genetic entity which could replicate either in the integrated or the independent state within the cell, and included in the class not only the F factor but those bacteriophages capable of existing in the prophage state integrated into the bacterial chromosome.

By 1953 a great deal was known about the bacterial nucleus and extensive chromosome maps had been made using classical genetic methods. The existence of the three methods by which bacteria can exchange genetic information was recognized in an increasing number of species. The mechanism of conjugation and recombination was being explored. It was believed that DNA must be the essential genetic material, but how it carried the information and how that information was conserved from generation to generation remained quite mysterious. It was at this point that the problem was suddenly advanced to the molecular level by the publication by Crick and Watson of their double helix model for DNA. The beauty of this model was not only that it gave a plausible three-dimensional structure and satisfied the thermodynamic requirements for the bond energies in the proposed molecule, but also that it postulated a semi-conservative mode of replication for the DNA macromolecule, each chain of the helix automatically generating its opposite number by selection from a pool of free nucleotides. Semi-conservative replication thus provided a possible answer to the riddle of the stability of the genome from generation to generation, and was eagerly accepted by biologists as the explanation.

There remained the problem of the nature of the genetic code. How could a macromolecule containing four and only four nucleotide bases code for the thousands of structural and phenotypic characters known to be passed on from parent to offspring in even the simplest of organisms? For some years geneticists had adhered to the belief that there was a one-to-one correspondence between genes and enzymes, 'one gene one enzyme' as the aphorism went, the gross structural and functional characters being readily understood to be produced as a result of the enzymes. Enzymes being proteins it was possible to redefine the problem in the form 'How can DNA consisting of only four different bases code for proteins which contain up to twenty different amino acids?' Eight years were consumed in the testing of one solution after another, each breaking down upon further testing, before in 1961 Crick, Barnett, Brenner and Watts-Tobin proposed the correct answer; that triplets of bases coded for individual amino acids. The code left a number of the sixty-four possible triplets unaccounted for, but this was no disadvantage as there must obviously be some way of coding instructions to start and stop reading the message and possibly for other directions. The correctness of the triplet code was soon verified by the elegant experiments of Niremberg using cell-free preparations from *E. coli* primed with synthetic polynucleotides. He first showed that the addition of polyuracil to the system led to the production of polyphenylalanine and then went on to study the effect of other artificial polynucleotides. His work was a brilliant confirmation of the hypothesis and it soon became apparent that the code was degenerate (i.e. more than one triplet codes for one amino acid) so that many more triplets were accounted for than had at first been thought, though even now there are a number of so-called 'amber triplets' whose value is not clear, but seem to be some sort of punctuation marks.

In 1963 the circular model of the chromosome of *E. coli* was confirmed by the autoradiographs published by Cairns, in which the structure could be seen exactly as predicted upon theoretical grounds.

A new field of bacterial genetics was opened up when in 1961 Watanabe (Fig. 8.4) published an account of the 'infectious drug resistance' first observed amongst shigellae in Japan on the late nineteen-fifties. This block transfer of resistance to several antibiotics from one species of enterobacteria to another was clearly of great clinical and public health importance. The mechanism was at first obscure, but over the last eleven years the

Fig. 8.4. Professor T. Watanabe (1923–
1972). From *Cellular Modification and Genetic
Transformation by Exogenous Nucleic Acids.
Sixth International Symposium on Molecular
Biology*, ed. R. F. Beers Jr and R. C. Tilgh-
man. © Johns Hopkins University Press
(1973).

work of Watanabe in Japan and of Datta, Meynell and Anderson
in the UK has made it clear that the resistance is mediated by one
or more types of 'conjugation promoting factors' similar to but
distinct from the F factors. The findings have shed light upon
the way that the F factor itself works. Electron micrographs have
demonstrated that each class of conjugation promoting factor is
associated with a specific pilus on the surface of the cells
possessing it and that these structures certainly cause the specific
adherence between donor and recipient cells and probably form
the conjugation tube through which the genetic material passes.
In addition a number of pilus-specific phages have been
discovered which attach to these organelles and are of considerable diagnostic value.

Bacterial and viral genetics continue to develop and in their
own right, but perhaps even more important is the new science
of molecular genetics to which they have given rise. Much of the
fundamental work continues to be done using the convenient

procaryotic systems, but during the last few years the exploration of the molecular genetics of the higher eucaryotic cells has
begun.

REFERENCES

1765 Spallanzani, L. *Saggo di Osservazioni Microscopiche Concernenti il Sistema della Generazione dei Signori di Needham e Buffon.* Modena. (French translation: *Nouvelles Recherches sur les Découvertes Microscopiques et la Génération des Corps Organises.* London and Paris: Lancombe (1769).)

1769 Needham, J. T. *Nouvelles Recherches Physiques et Metaphysiques sur la Nature et la Réligion avec une Nouvelle Théorie de la Terre, et une Mesure de la Hauteur des Alpes.* London and Paris: Lancombe.

1860 Pasteur, L. Experiences relatives aux generations dites spontanées. *C. r. Acad. Sci., Paris,* **50**, 303–7.

1907 Massini, R. Uber einen in biologischer Beziehung interessanten Kolis-tamm (*Bacterium coli mutabile*). *Arch. Hyg.,* **56**, 250–92.

1925 Mellon, R. M. Studies in microbic heredity. I. Observations on a primitive form of sexuality (zygospore formation) in the colon–typhoid group. *J. Bact.,* 481–501.

1927 Hadley, P. Microbic dissociation. *J. inf. Dis.,* **40**, 1–312.

1928 Griffith, F. The significance of pneumococcal types. *J. Hyg., Camb.,* **27**, 113–59.

1931 Dawson, M. H. & Sia, R. H. D. *In vitro* transformation of pneumococcal types. *J. exp. Med.,* **54**, 681–99.

1933 Alloway, J. L. Further observations on the use of pneumococcus extracts in effecting transformation of type *in vitro. J. exp. Med.,* **57**, 265–78.

1933 Reed, G. B. A hypothetical view of bacterial variation. *J. Bact.,* **25**, 580–6.

1941 Beadle, G. W. & Tatum, E. L. Genetic control of biochemical reactions in *Neurospora. Proc. nat. Acad. Sci. USA,* **27**, 499–506.

1942 Robinow, C. F. A study of the nuclear apparatus of bacteria. *Phil. Trans. R. Soc.,* ser. B, **130**, 299–324.

1943 Luria, S. E. & Delbruck, M. Mutations of bacteria from virus sensitivity to virus resistance. *Genetics,* **28**, 491–511.

1944 Avery, O. T., Macleod, C. M. & McCarty, M. Studies on the chemical nature of the substance inducing transformation of pneumococcal types. Induction of transformation by a deoxyribo-nucleic acid fraction isolated from pneumococcus type III. *J. exp. Med.,* **79**, 137–57.

1945 Tatum, E. L. X-ray induced mutant strains of *Escherichia coli. Proc. nat. Acad. Sci. USA,* **31**, 215–19.

1946 Hinshelwood, C. N. *The Chemical Kinetics of the Bacterial Cell.* London: Oxford University Press.

1947 Lederberg, J. Gene recombination and linked segregations in *Escherichia coli. Genetics,* **32**, 505–25.

1947 Weil, A. J. & Binder, M. Experimental type transformation of *Shigella paradysenteriae* (Flexner). *Proc. Soc. exp. Biol. Med.,* **66**, 349–52.

1947 Tatum, E. L. & Lederberg, J. Gene recombination in the bacterium *Escherichia coli. J. Bact.*, **53**, 673–84.

1948 Voureka, A. Sensitisation of penicillin-resistant bacteria. *Lancet*, **i**, 62–5.

1950 Tomscik, J. Uber eine bewegliche Mutante des *B. anthracis. Z. allg. Path. Bact.*, **13**, 616–24.

1951 Alexander, H. E. & Liedy, G. Determination of inherited traits of *H. influenzae* by desoxyribonucleic acid fractions isolated from type specific cells. *J. exp. Med.*, **93**, 345–59.

1951 McElroy, W. D. & Friedman, S. Gene recombination in luminous bacteria (*Achromobacter fischerii*). *J. Bact.*, **62**, 129–30.

1951 Lederberg, J. Prevalence of *Escherichia coli* strains exhibiting genetic recombination. *Science*, **114**, 68–9.

1951 Burnet, E. M. A genetic approach to variation in influenza viruses. (1, 2, 3 and 4). *J. gen. Microbiol.*, **5**, 46–82.

1952 Lederberg, J. & Lederberg, E. M. Replica plating and indirect selection of bacterial mutants. *J. Bact.*, **63**, 399–406.

1952 Zinder, N. D. & Lederberg, J. Genetic exchange in *Salmonella. J. Bact.*, **64**, 679–99.

1952 Hayes, W. Recombination in Bact. *E. coli* K12: unidirectional transfer of genetic material. *Nature, Lond.*, **169**, 118–19.

1952 Lederberg, J., Cavalli, L. L. & Lederberg, E. M. Sex compatibility in *Escherichia coli. Genetics*, **37**, 720–30.

1953 Hayes, W. The mechanism of genetic recombination in *Escherichia coli. Cold Spring Harb. Symp. quant. Biol.*, **18**, 75–93.

1953 Visconti, N. & Delbruck, M. The mechanism of genetic recombination in phage. *Genetics*, **38**, 5–33.

1953 Crick, F. H. C. & Watson, J. D. Molecular structure of nucleic acids. A structure for deoxyribose nucleic acid. *Nature, Lond.*, **171**, 737–8.

1954 Delamater, E. D. Cytology of bacteria. II. The bacterial nucleus. *Ann. Rev. Microbiol.*, **8**, 23–46.

1955 Holloway, B. W. Genetic recombination in *Pseudomonas aeruginosa. J. gen. Microbiol.*, **13**, 572–81.

1956 Robinow, C. F. A study of the nuclear apparatus of bacteria. *Bact. Rev.*, **20**, 207–92.

1956 Belser, W. L. & Bunting, M. I. Studies on a mechanism providing for genetic transfer in *Serretia marcasens. J. Bact.*, **72**, 582–92.

1956 Yura, T. Evidence of non-identical alleles in purine-requiring mutants of *Salmonella typhimurium*. In *Genetic Studies with Bacteria*, ed. M. Demerec *et al.*, pp. 63–75. Washington: Carnegie Institute.

1957 Wollman, E. L. & Jacob, F. Sur les processus de conjugaison et de recombinaison chez *Escherichia coli. Ann. Inst. Pasteur*, **93**, 323–39.

1957 Luria, S. E. & Burrous, J. W. Hybridization between *Escherichia coli* and *Shigella. J. Bact.*, **74**, 461–76.

1959 Timakov, L. *Microbial Variation*. Oxford: Pergamon Press.

1961 Crick, F. H. C., Barnett, L., Brenner, S. & Watts-Tobin, R. J. General nature of the genetic code for proteins. *Nature, Lond.*, **192**, 1227–32.

1961 Watanabe, T. & Fukasawa, T. Episome mediated transfer of drug resistance in enterobacteriaceae. *J. Bact.*, **81**, 669–83.

1962 Datta, N. Transmissible drug resistance in an epidemic strain of *Salmonella typhimurium*. *J. Hyg., Camb.*, **60**, 301–10.

1963 Cairns, J. The chromosome of *Escherichia coli. Cold Spring Harb. Symp. quant. Biol.*, **28**, 43–6.

1965 Anderson, E. S. & Datta, N. Resistance to penicillins and its transfer in enterobacteriaceae. *Lancet*, **i**, 407–9.

1965 Meynell, E. & Datta, N. Functional homology of the sex factor and resistance transfer factors. *Nature, Lond.*, **207**, 884–5.

9

Serology and immunology

The study of immunology commenced with observations upon the whole animal. It then passed through a phase when interest was centred upon the antibacterial properties of blood plasma and the phagocytic cells of the body, and the question as to which of these two mechanisms was the effective means by which immunity to bacterial and viral infections was achieved. Finally after the importance of both the cellular and the humoral defences was recognized, interest centred upon the physico-chemical nature of antigen–antibody reactions and the mechanisms responsible for the synthesis of specific antibodies.

The fact that attacks of certain infectious diseases are followed by immunity to further illness of the same kind has been known since antiquity. The artificial infection of men and animals to protect them against infections such as smallpox and rinderpest has been practised as an empirical procedure since classical times or earlier. Jenner's 'vaccination' against the smallpox was but an extension of this age-old procedure, without any understanding of the mechanisms involved.

The first rational attempt to produce artificial active immunity was made in 1880 by Louis Pasteur. The story is a very good example of how an apparent failure in a series of experiments was, by the genius of this great scientist, made the starting point for a whole new field of preventive medicine. Pasteur was investigating the pathology of fowl cholera, a disease that we now know to be caused by infection with *Pasteurella septica*. By the summer of that year he had obtained the causative organism in pure culture and had succeeded in inducing the disease in healthy fowls by injections of such cultures. At this point in the work the long vacation intervened and Pasteur placed his broth cultures in a cupboard, shut up his laboratory and departed with his family for a holiday in the country. When he returned to Paris in September his first step was to inject some of the broth cultures that he had preserved over the vacation into fowls in order to establish further experimental infections. To his surprise the birds were unaffected and remained well in spite of injections comparable to those that had killed similar animals in the summer. Faced with this apparent failure Pasteur embarked upon a fourfold experiment. While continuing to maintain the laboratory strain of the causative organism, he re-isolated a fresh strain from birds suffering from a natural infection. He also preserved the birds that he had unsuccessfully challenged. In addition he purchased a new stock of fowls from the market. He now proceeded to inject the fresh cultures into both the old stock of hens and the new stock, and likewise the old culture into both types of fowl. To his surprise he found that the hens which had been previously given the old culture were resistant to challenge with the fresh cultures, while the new batch of fowls became ill and died when injected with this type of culture. The old cultures failed to produce illness in either group of fowls. He thus proved that the difference in results between the summer and autumn experiments was due to an alteration in the virulence of the bacterial cultures and not to some difference in the fowls used in the experiments. He had further shown that injection with the attenuated strain of bacterium conferred upon the birds resistance to subsequent challenge with fully virulent organisms, capable of killing normal animals. Pasteur went on to investigate the factors involved in this loss of virulence and showed that it was related to the length of time between subcultures; the longer the interval the greater the degree of attenuation. In the paper in which he published his findings he drew attention to the analogy with the procedure of vaccination introduced by Jenner nearly ninety years before. He wrote:

It seems from the above facts that in fowl cholera, there exists a state of the virus relative to the most virulent virus, which acts in the same way as cowpox virus does in relation to smallpox virus. Cowpox virus brings about a benign illness, cowpox, which immunizes against a very serious illness, smallpox. In the same way, the fowl cholera virus can occur in a state of virulence that is sufficiently attenuated, so that it induces the disease but does not bring about death, and in such a way that after recovery, the animals can survive an inoculation with the most virulent virus. Nevertheless, the difference between smallpox and fowl cholera is considerable in certain respects, and it is not amiss to remark that, with respect to an understanding of the principles, studies on fowl cholera will probably be more helpful. Whereas there is still a dispute about the relationship between smallpox and cowpox, we know for certain that the attenuated virus of fowl cholera is derived directly from the most virulent virus of this disease, so that their natures are fundamentally the same.

Pasteur referred to the live attenuated strains of the fowl cholera organisms as vaccines, and the term has continued to be used to this day. In the next year Pasteur went on to apply the principles of vaccination with an attenuated strain of the causative organism to the case of anthrax. In this instance the clue to the method of attenuation came, not by chance, but by a consideration of the comparative pathology of the disease. It was known that birds do not suffer from natural infections with the anthrax bacillus, and Pasteur showed that they are resistant to experimental infections under normal conditions. It occurred to him that the resistance of birds might be due to their high body temperature, 44 °C, and to test this hypothesis he injected anthrax bacilli into a hen whose body temperature was artificially lowered by placing it in a bucket packed round with ice. Such hypothermic birds were found to be susceptible to the infection, dying in a typical septicaemic state. Further experiments showed that anthrax bacilli grown at 44 °C in the test-tube lacked a capsule and were of reduced virulence. These attenuated cultures were the basis of the first anthrax vaccine. Laboratory experiments having indicated that the vaccine conferred a considerable degree of protection against challenge with fully virulent organisms, the famous field trial at Pouilly le Fort was carried out with complete success. This trial attracted so much interest that the famous *Times* correspondent De Blowitz wrote a special report of it for the paper.

In 1882 the first observations on cellular immunity were made by the Russian biologist Metchnikoff (Fig. 9.1). He observed the

Fig. 9.1. Elie Metchnikoff (1845–1916). From an original photograph in the Wellcome Historical Medical Museum. By courtesy of 'The Wellcome Trustees'.

collection of amoeboid cells about a rose thorn that he had introduced into the translucent body of a starfish larva. He coined the name phagocytes for these cells which devoured the foreign material, and over the next thirty years he collected a vast mass of observational data in support of his hypothesis that resistance to infection was due entirely to the phagocytic cells and not to the circulating antibodies: the cellular hypothesis of

Fig. 9.2. Theobald Smith (1859–1934).

immunity as opposed to the humoral. The controversy was ultimately proved to be sterile, but while it continued it gave rise to a great deal of valuable experimental data and advanced our knowledge of the body defences against infection.

In 1884 Louis Pasteur commenced his classical studies on rabies which led in 1885 to the development of a vaccine against this disease. Perhaps the most significant aspect of the work was the way that Pasteur was able, by applying the general principles that had emerged from his earlier work on fowl cholera and anthrax, to produce an attenuated form of a virus which he could not either see under the microscope or cultivate *in vitro*. In spite of these obstacles he succeeded in producing the first effective treatment for this otherwise uniformly fatal disease. The international acclaim that followed this work led to the setting up of the Institut Pasteur and a great increase in the support for medical microbiology all over the world. The intellectual achievement of Pasteur was matched only by the personal courage required to operate day after day with rabid dogs, well knowing that a single bite or scratch would be followed by a certain and agonizing death. Working with rabies is a hazardous and unpleasant business today, but Pasteur's early work before he succeeded in developing the vaccine must have required a combination of technical brilliance and personal heroism seldom equalled in the history of science.

Up until 1885 all vaccines that had been developed were live attenuated preparations. In 1886 the first killed vaccine was produced by the American veterinary bacteriologists Salmon and Theobald Smith (Fig. 9.2). Hog cholera, caused by a member of the genus *Salmonella*, was a serious economic problem in the middle west of the USA, the so-called 'hog belt'. The killed vaccine was developed as part of a programme for the control of this disease and was to be as effective in this infection as the live vaccine had been against fowl cholera. The discovery that active immunity could be induced by the injection of killed organisms was a very important step forward. Not all bacteria and viruses can be attenuated, and even when this is possible the production of a standardized live vaccine is a very expensive and time-consuming affair. Killed, fully virulent organisms provided a type of vaccine that was much simpler to produce and standardize and in addition could be stored for longer periods without deterioration and was thus simpler to use in the field. From Salmon and Smith's work a host of killed vaccines effective against various human and animal infections have since been developed.

Now that practical methods were available for producing artificial active immunity in men and animals, bacteriologists began to study the mechanisms by which this immunity was mediated. It had been known for some years that blood was relatively resistant to putrefaction, and it was obviously of interest to study the effect of the serum independently of the cells upon the growth of bacteria in the test-tube. The earliest studies published were those of the English bacteriologist Nuttall in 1888. He added known numbers of anthrax bacilli to cell-free serum and noted that so long as the number added was not too great the bacteria were killed by the serum. There are two points that should be noted about this early work on the bactericidal effect of serum: in the first place the serum was from non-immune animals and the effects were regarded as non-specific property of the serum or plasma; in the second place the technique of measuring the bactericidal effect of samples of serum was the opposite to present day methods. The quantity of serum was held constant and the number of bacteria added to it was varied, so that the results were expressed as the number of bacteria killed by a unit volume of serum, not as a limiting dilution of serum capable of producing the death of a fixed number of bacteria.

Nuttall's observations were confirmed and extended to other species of bacteria by Büchner in the following year (1889). In the same year the first evidence that immune serum differed from non-immune serum in its antibacterial properties was produced by the Frenchmen Charrin and Roger. They found that when *Pseudomonas pyocyanea* was grown in the presence of serum from animals that had recovered from artificial infections with this organism, the bacilli formed clumps which fell to the bottom of the test-tube. Growth in the serum of animals that had never been infected with *P. pyocyanea* was in contrast diffuse. This was the first evidence for specific antibacterial substances in the sera of immune animals.

The next decade, the eighteen-nineties, was a period of great growth for serology and immunology. The existence of specific antibacterial substances produced by the host following infection was confirmed, antitoxic immunity was discovered and exploited in the treatment and prevention of tetanus and diphtheria, the complex phenomenon of specific lysis unravelled and the first diagnostic serological test brought into use in clinical medicine.

During the year 1890 Von Behring (Fig. 9.3) dominated the scene. Early in the year he made the first report of acquired specific immune antibodies as a result of his studies on experimental infections of guinea-pigs with *Vibrio metchnikovi*. Later, in collaboration with the Japanese bacteriologist Kitasato, he induced artificial active immunity to tetanus and diphtheria toxin in the guinea-pig and then went on to show that this immunity could be transferred passively to other animals by the injection of serum from an immune animal, thus laying the foundation of serotherapy. Von Behring and Kitasato were also responsible for an important theoretical advance. They it was who coined the word 'antitoxin' for the specific substance in the serum of immune animals which could confer immunity upon normal animals when injected into them. It was from their

Fig. 9.3. Emil A. Von Behring (1854–1917).
From *Parasitology* (1924), **16**.

suggestion that the more general concept of a class of antibodies was to emerge later. It was only just over a year after the publication of Von Behring and Kitasato's paper that the first child was treated with anti-diphtheriatic serum by Geisslin of Berlin, on Christmas night in 1891.

It was in 1891 also that Paul Ehrlich published his first paper on the subject of immunology. For the next ten years Ehrlich was the leading figure in the field, devising the quantitative techniques that were essential for the production of standardized sera for clinical use, producing the most successful model of specific lysis and propounding a theory of antibody production that dominated men's thinking on this subject for the next thirty years. If we are to understand Ehrlich's contributions to immunology we must see them against the background of his earlier work upon the specific staining of blood cells and the theoretical concepts set forth in his monograph *The Oxygen*

Requirements of the Organism, which led him to his successful treatment of malaria with the dye methylene blue. Terms such as affinity, avidity and receptor site, were all taken over from the chemical theory of dyeing, applied first to the staining of cells and then to the action and production of antibodies. Perhaps because he was trained as a physician and not as a chemist Ehrlich used these chemical concepts in a very imaginative way and was never concerned with their rigorous application nor with any attempt to interpret the biological phenomena in terms of physical and chemical realities.

Ehrlich's paper of 1891, in which he differentiated clearly between active and passive immunity and drew general conclusions from his observations, is important as exemplifying the value of simplified artificial systems in biological research. Von Behring and Kitasato working with the natural infections tetanus and diphtheria had observed that it was possible to transfer immunity from an immune animal to a non-immune one by injections of serum and had pointed out the therapeutic possibilities, but they had not drawn the general conclusion that here are two types of immunity; active and passive. Ehrlich worked with the vegetable poison ricin which, when injected into experimental animals, causes death as a result of an acute haemolytic anaemia. He was able by starting with minute doses of the poison and gradually increasing them day after day to induce a state of immunity in animals such that they were unaffected by the injection of many lethal doses of the material. He then showed that the sera of these immune animals when injected into normal animals conferred protection against challenge with lethal doses of the poison. As he was working with a chemically defined poison he was able to make quantitative studies and determine how many milligrams of ricin one millilitre of serum from animals with different degrees of induced immunity would neutralize. He further, and this was the most important concept of the paper, distinguished between active and passive immunity. The special case of tetanus or diphtheria was generalized.

It was Büchner in 1893 who first showed that the bactericidal property of serum was lost after the serum had been heated to 56 °C. He coined the term 'alexine' for the heat-labile substance that he had discovered. Büchner appears to have thought that the bactericidal properties of serum were entirely dependent upon the presence of alexine. This was possible because he was concerned with general rather than specific bactericidal effects.

An understanding of the nature and properties of the heat-labile component involved in the bactericidal action of serum came from an extension of the work of Pfeiffer on the lysis of cholera vibrios in the peritoneal cavity of actively immunized guinea-pigs. In 1894 Pfeiffer reported that the effect was produced by immune serum *in vitro*, and later that same year Fraenkel and Sobernheim showed that if the serum was heated to 70 °C for one hour the bacteriolytic effect was lost in the test-tube, but curiously, injection of heat-inactivated serum into normal guinea-pigs, conferred passive immunity upon them so that when cholera vibrios were injected into their peritoneal cavities they were lysed as rapidly as if the animals had been actively immune. Metchnikoff carried the investigation a step further in 1895 when he showed that specific bacteriolysis of cholera vibrios would take place in the test-tube in the presence of heated immune serum provided that fresh peritoneal exudate was added to the system. Later in the same year the non-specific nature of the heat-labile substance was established by Jules Bordet who was working at the time in Metchnikoff's laboratory at the Institut Pasteur; he showed that the power of lysing cholera vibrios in the test-tube was restored to heated immune serum by the addition of normal fresh unheated serum. He showed further that the critical temperature at which the heat-labile substance was destroyed in any serum was 56 °C for one hour.

During the next four years no major advances were made in the understanding of the complex mechanism of bacteriolysis, but other properties of immune serum were investigated and turned to diagnostic use in medicine. During the years 1894 and 1895 Durham, an English bacteriologist working as a post-graduate student in Von Gruber's laboratory in Vienna (Fig. 9.4), followed the hint given by Charrin and Roger's paper of 1889 and made a thorough study of the agglutinating property of immune sera. He showed that the agglutination was specific, made quantitative studies using fixed antibody and variable antigen concentrations, and pointed out how the phenomenon of specific agglutination could be used diagnostically both to

Fig. 9.4. Max von Gruber (1853–1927). From
Münch. Med. Wochenschr. (1964) **106**, p. 93.

identify unknown organisms, such as possible typhoid bacilli,
and to establish a diagnosis of typhoid fever by testing the power
of the patient's serum to agglutinate known suspensions of
typhoid bacilli. It was unfortunate that the publication of this
work was delayed until 1897, two years after it had been
completed. In the meanwhile, in 1896 the French bacteriologist
Widal published an account of the use of the agglutination
reaction as a diagnostic aid in cases of suspected typhoid fever.
Thus the test is known to this day as the 'Widal reaction'
although Durham's work antedates that of Widal by at least a
year.

It was in the following year, 1897, that Kraus published the
first account of precipitation reactions when immune sera were
added to cell-free filtrates of homologous bacterial cultures. The
organisms that he worked with were cholera, typhoid and
plague bacilli.

In this year also Paul Ehrlich published his classical paper on 'The standardization of diphtheria antiserum and its theoretical basis'. In this article he laid down the basis for all future quantitative work on toxins and antitoxins. Having first shown that diphtheria toxin was unstable in storage, losing its toxic property while retaining its combining power with antitoxin, he went on to state that the fundamental unit must be one based upon a preserved standard specimen of antitoxin. With this material it was possible to standardize any batch of toxin and then use this to standardize further batches of serum, provided that the tests were carried out within a day or two. His results proved that the neutralization of toxin by antitoxin was not a simple affair comparable to the reaction between a strong acid and a strong alkali, and he was obliged to propose two new quantitative concepts. These were the L_0 (Limes nul) dose of toxin, defined as 'the largest amount of toxin which, when mixed with one unit of antitoxin and injected subcutaneously into a guinea-pig of 250 grams weight gives no reaction', and the $L+$ (Limes tod) dose defined as 'the smallest amount of toxin which when mixed with one unit of antitoxin and injected subcutaneously into a 250-gram guinea-pig will kill the animal within 96 hours'. Ehrlich originally defined the unit of antitoxin as that amount which would neutralize 100 MLD (minimum lethal doses) of a particular batch of toxin that he had produced in his laboratory; the MLD he had defined as the smallest amount of toxin that when injected subcutaneously into a 250-gram guinea-pig would kill the animal within 96 hours. However, once the instability of toxin was appreciated the definition of a unit of antitoxin was altered to read 'that amount of serum that has the same total combining power for toxin and toxoid together as one unit of Ehrlich's original antitoxin'. Toxoid was the term coined to describe the degraded toxin which had lost its toxicity but still retained its combining power. These new units together with the statistical techniques which Ehrlich introduced to deal with the biological variation in susceptibility to toxin found amongst guinea-pigs made it possible to lay down standard test procedures which enabled antisera of comparable potencies to be prepared not only at different times in the same laboratory but in different laboratories. The present day national and international stan-

Fig. 9.5. Jules Bordet (1870–1961)..
From A. Mansch, *Medical World.* Berlin-
Charlottenburg, Adolf Eckstein (1906).

dards for antitoxic sera are all extensions of Ehrlich's original
proposals.

The next step in the understanding of the lytic properties of
immune serum was made by Jules Bordet (Fig. 9.5). This is
another example of the value of a simplified artificial model.
The earlier work on bacteriolysis had all been read by microscopic examination of suspensions of vibrios, a slow and labori-
ous method which limited the rate of progress of the work. In
1898 Bordet published a paper describing results obtained using
another system, the lysis of the red blood cells of one species of
animal by immune sera raised against them in second species by
injection of the cells of the first species. This specific haemolysis
could be read with the naked eye, as the lysis of the red blood

cells released haemoglobin into the liquid phase of the system, so that after standing on the bench for a time tubes in which lysis had taken place were uniformly red while tubes where lysis had not taken place had a clear supernatant and a deposit of intact erythrocytes. The haemolytic system had the further advantage that it was not complicated by any relation to infection and defence mechanisms. It could be studied in its own right as a physiological reaction. Bordet's studies using this system showed that specific haemolysis, like bacteriolysis, was effected only when two substances were present together in the serum; a heat-labile substance present to a greater or lesser degree in all fresh sera, and heat-stable specific acquired antibodies. He showed further that the non-specific heat-labile factor could be supplied by fresh serum from an animal of a different species from the one in which the specific immune serum had been raised. Guinea-pig serum proved to be a particularly rich source of the non-specific heat-labile factor. The system used by Bordet and others for much of the later work consisted of a rabbit anti-sheep cell serum inactivated by heat and reactivated by the addition of fresh guinea-pig serum. In the 1898 paper the studies were made using guinea-pig anti-rabbit sera reactivated after heating with fresh non-immune guinea-pig serum. The results showed that the non-specific heat-labile substance was essential for lysis of the red cells by the specific immune serum, but not for agglutination of the red cells by the specific immune serum. Heat-inactivated immune serum had lost its haemolytic power but retained its ability to haemagglutinate. Bordet proposed a model to explain his finding. He visualized the heat-labile non-specific substance as acting upon red cells which were sensitized by the action of the specific haemolytic serum. He named the heat-stable component of the system 'substance sensibilitrice' and the heat-labile component 'alexine', the term coined by Büchner, and ascribed the phenomenon to the destruction of a hypothetical substance of that name. Bordet's model of haemolysis did not involve any union between the substance sensibilitrice and the heat-labile alexine. The alexine was able to lyse the red blood cell only after its surface structure had been modified by the action of the specific substance sensibilitrice, but that was the only relation between the two substances. Bordet's model was challenged in 1900 by an alternative scheme advanced by Paul Ehrlich: in this model the heat-labile factor was referred to as 'complement' and the heat-stable specific factor as 'amboceptor'. More important than the changed terminology is the different roles allotted to the two substances in Ehrlich's scheme. He visualized the specific heat-stable amboceptor as possessing two haptophore groups, the one cytophilic which attached to the red cell and the other complementophilic which attached to one end of the complement molecule. This union enabled the complement molecule to cause the lysis of the red cell in a manner that was not made clear in the hypothesis. Ehrlich's hypothesis and the associated terminology were more popular amongst the bacteriologists of the time and came after a few years to be the accepted model.

It was Bordet and Gengou who in 1901 first applied the new knowledge of specific haemolysis to the detection of other antigen – antibody reactions. They showed that while complement was only necessary for lysis to take place, it was used up whenever any antigen–antibody reaction took place. They therefore developed the complement fixation test for the detection of antibody–antigen reactions by using a complement-free haemolytic system (sensitized red cells) as an indicator. Their first paper dealt with complement fixation during specific lysis of cholera vibrios. Pfeiffer's original technique, depending as it did upon microscopic detection of lysis of the bacterial cells, was extremely tedious; Bordet and Gengou's technique enabled the results to be read at a glance by inspection of the tubes after addition of the indicator haemolytic system and re-incubation. The technique has since been applied to a number of diagnostic problems, the best known application being the Wassermann test for antibodies in syphilis. Today complement fixation tests are perhaps most widely used in virology.

In the same year (1901) Landsteiner (Fig. 9.6) published the discovery of the natural antibodies against red blood cells of the same species in man and by the use of these was able to establish the existence of the four ABO blood groups. This was not only a discovery of great practical importance, making blood transfusion a therapeutic possibility, but it was also of fundamental biological significance. For the first time it was shown that antibodies were not produced only as a defence against infection or the introduction of the cells of animals of another species into

Fig. 9.6. Karl Landsteiner (1868–1943). From MacCullum & Taylor, *The Nobel Prize Winners, 1901–1937*. Zurich, The Central European Times Publishing Co. Ltd (1938).

the body of the host, but played a vital role in what came later to be referred to as self-recognition. The implications of Landsteiner's discovery for future work on the origin and production of antibodies was not immediately appreciated. There is little evidence for instance that Ehrlich felt obliged to modify his monumental side chain theory to account for the new facts, but as the century wore on more and more significance was attached to the results.

The year 1902 may be regarded as the beginning of immunopathology. In that year Portier and Richet described the phenomenon of anaphylaxis in dogs injected with extract of sea urchins. It soon became clear that this perversion of the immune response was of widespread occurrence, differing in its manifestation from species to species but dependent in all cases upon the same basic mechanism. This mechanism was elucidated by Dale

Fig. 9.7. Sir Almroth Wright (1861–1947).
Reproduced by kind permission of the Royal
Army Medical College.

and his co-workers in a series of classic papers published from 1913 onwards. By the mid nineteen-twenties it had been established that the release of histamine following damage to cells by combination between antigen and fixed antibodies on their surfaces accounted for most of the observed facts; however there still remain areas of darkness in our picture.

In 1903 in a contribution to the *Comptes rendues des séances de la Société de biologie*, Arthus drew attention to the phenomenon to which his name is attached. This was another perversion of the immune response. Both the Arthus phenomenon and experimental anaphylaxis were admittedly seen only under artificial conditions in the laboratory, but they served to draw attention to the fact that the immune response was not an unmixed blessing, and could under certain circumstances prove fatal to the host.

Fig. 9.8. Captain S. R. Douglas (1871–1936).
From *Obit. Not. Fellows R. Soc.* (1936), **2**,
p. 175.

The relevance of this line of thought was underlined when in 1905 Von Pirquet and Schick published their monograph on 'serum sickness'. They described a condition that was of obvious clinical importance, and one that only occurred following serotherapy; a new iatrogenic disease that was bound to become more frequent as antitoxic and antibacterial sera were more widely used in treatment and prophylaxis.

In the year 1903 Wright (Fig. 9.7) and Douglas (Fig. 9.8) published a paper which effectively ended the twenty-years-long controversy between the protagonists of the cellular theory of immunity and their opponents who believed in the humoral theory. With the discovery of opsonins, antibodies that acted by enhancing the phagocytic activity of the polymorphonuclear leucocytes, it became clear that the two mechanisms were complementary and discussion in terms of either/or was meaningless.

The application of the complement fixation test to the diagnosis of syphilis by Wassermann and his colleagues in 1906 was not only an important step forward in diagnosis but also an interesting story in the history of science. The Austrian workers at first used as an antigen the livers of congenitally syphilitic foetuses as it was known that these organs contained vast numbers of treponemata and might therefore be presumed to be a good source of specific antigen. French workers then repeated the tests using extracts of normal livers and found that they still got positive tests only in cases that were clinically known to be suffering from syphilis. This led the group to Vienna to test other extracts as an antigen for the test and ultimately to settle upon the cardiolipid antigen prepared from bovine heart muscle, which with modifications is still in use today. It is presumed that there must be shared antigenic determinants between the bovine heart muscle and the spirochaete, but the theoretical basis of the test is still not understood in spite of its worldwide use over the last sixty years.

An important step forward in our understanding of the nature of antibodies was made when in 1911 Landsteiner showed that antibody activity was associated with the globulin fraction of the serum. This was the beginning of the chemical characterization of these substances first postulated as theoretical entities to explain the phenomena of agglutination, precipitation and bacteriolysis which occurred in the presence of immune sera. The further chemical analysis of these substances was to take over twenty years, but now we have not only a clear idea of the types of compounds that show antibody activity, but even of their molecular structures, and work is at present going on to determine the chemical basis of their specificity.

From 1917 onwards Landsteiner and his group devoted themselves to the study of the chemical basis of the specificity of antigens. Working with synthetic antigens made by coupling small organic molecules of known structure to proteins such as egg albumin they were able to observe the effect of substitution of one group by another and in the case of benzene derivatives the effect of substituting the *ortho* for the *meta* or *para* form of the molecule. They found that even such slight alterations of the determinant molecule altered the specificity of the antigen, as measured by its reactivity with specific sera. It was thus established that antigenic specificity was explicable in strictly chemical terms and the way opened for a more rigorous theory of antigen–antibody reactions than was provided by Ehrlich's famous side chain theory, which by the time of his death in 1915 had become so encrusted with theoretical entities designed to deal with specific objections that it began to appear more as a brilliant piece of special pleading than a genuine scientific hypothesis. Landsteiner's results also made available to biologists and chemists a new and extremely sensitive analytical technique. It was already appreciated that serological analysis could differentiate between the proteins of different species of animals and even between subgroups within a species, as in the case of the human blood groups: it was now clear that it could be used to detect differences of single atoms at molecular level.

One important result of the chemical approach to the specificity of antigens was the discovery by Heidelberger and Avery in 1923 that the specific capsular antigens of pneumococci were high molecular weight polysaccharides. Up to this time there had been a general consensus that only proteins were able to act as full antigens. It was now clear that polysaccharides, if of sufficient molecular weight, could also provoke the formation of antibodies.

During the nineteen-twenties and nineteen-thirties the interest of serologists centred upon the problems of how antigen–antibody reactions took place. Earlier there had been two schools of thought about this matter; one which explained the reactions in strictly chemical terms and supported by Ehrlich, and the other which explained the phenomena in physical terms as reactions between colloids, originating from the ideas of Bordet. The quantitative studies of Dean and other workers on the precipitin reaction and the recognition of the importance of the optimal proportion between antigen and antibody led to refinements in the theories put forward to explain the antigen–antibody reactions. For some years the question of the valency of antibody was debated, but finally in the nineteen-thirties with the publication of Marrack's lattice hypothesis it became generally agreed that they must be multivalent and were probably bivalent.

It was during the nineteen-thirties that certain physical chemists became interested in the problem of serology. The two greatest were Linus Pauling and Tiselius; between them they altered the whole attitude of serologists to their problems, defining them in concrete physico-chemical terms which led to hypotheses that were testable by quantitative measurements, and introducing new and powerful analytical tools into the laboratory.

Serology since 1939 has been dominated by the use of three new physical techniques. First electrophoresis, introduced for the separation of globulins by Tiseleus in 1939; then the application of chromatography which together with analytical ultracentrifugation made possible the determination of the molecular weights of antibodies and their component polypeptide chains; and thirdly the use of electron microscopy to study the reactions between antibodies and antigens labelled with heavy atoms. Progress has been steady and we now have a coherent picture of the various types of antibody, some monomers and some polymers, and all built up from light and heavy polypeptide chains. The application of the 'finger-printing' (Porter) technique, so successful in unravelling the amino acid sequence of the haemoglobins, is beginning to provide the information that is needed to understand the molecular basis of the specificity of the antibodies.

REFERENCES

1880 Pasteur, L. De l'atténuation du virus du choléra des poules. *C. r. Séanc. Acad. Sci.*, **91**, 673–80.

1881 Pasteur, L. De l'atténuation des virus et de leur retour à la virulence. *C. r. Séanc. Acad. Sci.*, **92**, 429–35.

1881 Pasteur, L. Le vaccin du charbon. *C. r. Séanc. Acad. Sci.*, **92**, 666–8.

1884 Metchnikoff, E. Uber eine Sprosspilzkrankheit der Daphnien. Beiträge zur Lehre über den Kampf der Phagocyten gegen Krankheitserreger. *Virchows Arch. path. Anat. Physiol.*, **96**, 177–95.

1885 Pasteur, L. Méthode pour prevenir la rage après morsure. *C. r. Séanc. Acad. Sci.*, **101**, 765–73.

1886 Salmon, D. E. & Smith, T. The bacterium of swine-plague. *Am. mon. micr. J.*, **7**, 204–5.

1888 Nuttall, G. Experimente über die bakterienfeindlichen Einflusse des thierischen Körpers. *Z. Hyg. InfektKrankh.*, **4**, 353–94.

1889 Büchner, H. Ueber die Bakterientodtende Wirbung das zellenfreien Blutseriens. *Zentralbl. Bakt.*, **5**, 817–23.

1889 Charrin, A. & Roger, J. Note sur le développement des microbes pathogèns dans le sérum des animaux vaccinés. *C. r. Séanc. Soc. Biol.*, **1**, 667–9.

1890 Behring, E. von & Nissen, F. Ueber bakterienfeindliche Eigenschaften verschiedener Blutserumarten. *Z. Hyg. InfektKrankh.*, **8**, 412–33.

1890 Behring, E. von & Kitasato, S. Ueber das Zustandekommen der Diphtherie-Immunität und der Tetanus-Immunität bei Thieren. *Dtsch. Med. Wochenschr.*, **16**, 1113–14.

1891 Ehrlich, P. Experimentelle Untersuchungen über Immunität. *Dtsch. med. Wochenschr.*, **17**, 976–9 (I: Ueber Ricin), 1218–19 (II: Ueber Abrin).

1893 Büchner, H. Ueber Bakteriengifte und Gegengifte. *Münch. med. Wochenschr.*, **40**, 449–52.

1894 Fraenkel, C. & Sobernheim, G. Versuche über das Zustande kommen der kunstlichen Immunität. *Hyg. Rundschau. Berl.*, **4**, 97, 145.

1894 Pfeiffer, R. Weitere Untersuchungen über das Wesen der Choleraimmunität und über specifische baktericide Processe. *Z. Hyg. InfektKrankh.*, **18**, 1–16.

1894 Roux, M. E., Martin, M. L. & Chaillou, M. A. Trois cents cas de diphtérie traites par le serum antidiphtérique. *Ann. Inst. Pasteur*, **8**, 640–61.

1895 Metchnikoff, E. Sur la destruction extracellulaire des bacteries dans l'organisme. *Ann. Inst. Pasteur*, **9**, 433–61.

1895 Bordet, J. Les leucocytes et les propriétés actives du sérum chez les vaccinés. *Ann. Inst. Pasteur*, **9**, 462–506.

1896 Gruber, M. von. Theorie der aktiven und passiven Immunität gegen Cholera, Typhus, und verwandte Krankheitsprocesse. *Münch. med. Wochenschr.*, **43**, 206–7.

1896 Widal, G. F. I. & Sicard, A. Recherches de la réaction agglutinante dans le sang et le sérum desséchés des typhiques et dans la sérosité des vésicatoires. *Bull. Mém. Soc. méd. Hôp., Paris*, 3ᵉ sér., **13**, 681–2.

1897 Gruber, M. A theory of active and passive immunity from the bacteria of cholera, typhoid fever and the like. *Lancet*, **ii**, 910–11.

1897 Kraus, R. Ueber specifische Reaktionen in keimfreien Filtraten aus Cholera, Typhus und Pest-bouillonculturen, erzeugt durch homologes Serum. *Wien. klin. Wochenschr.*, **10**, 736–8.

1897 Durham, H. E. On a special action of the serum of highly immunised animals. *J. Path. Bact.*, **4**, 13–44.

1897 Ehrlich, P. Die Werthbemessung des Diphtheriaheilserums und deren theoretische Grundlagen. *Klin. Jahrb.*, **6**, 299–326.

1897 Ehrlich, P. Zur Kenntniss der Antitoxinwirkung. *Fortschr. Med.*, **15**, 41–3.

1898 Bordet, J. Sur l'agglutination et le dissolution des globules rouges par le sérum d'animaux injectés de sang défibrine. *Ann. Inst. Pasteur*, **12**, 688–95.

1900 Ehrlich, P. & Morgenroth, J. Ueber Hämolysine. *Berl. klin. Wochenschr.*, **37**, (1), 453–50, (11), 681–7.

1901 Bordet, J. & Gengou, O. Sur l'existence de substances sensibilitrices dans la plupart des sérums antimicrobiens. *Ann. Inst. Pasteur*, **15**, 289–302.

1901 Bordet, J. Sur la mode d'action des sérums cytolytiques et sur l'unité de l'alexine dans un même sérum. *Ann. Inst. Pasteur*, **15**, 303–18.

1901 Landsteiner, K. Ueber Agglutinationserscheinungen normalen menschlichen Blutes. *Wien. klin. Wochenschr.*, **14**, 1132–4.

1902 Portier, P. & Richet, C. De l'action anaphylactique de certains venins.
C. r. Séanc. Soc. Biol., **54**, 170–2.

1903 Wright, A. E. & Douglas, S. R. An experimental investigation of the role
of the blood fluids in conection with phagocytosis. *Proc. R. Soc.*, **72**,
357–70.

1906 Wassermann, A., Neisser, A. & Bruck, C. Eine serodiagnostische
Reaktion bei Syphilis. *Dtsch. med. Wochenschr.*, **32**, 745–6.

1923 Heidelberger, M. & Avery, O. T. The soluble specific substance of
Pneumococcus. J. exp. Med., **38**, 73–9.

1930 Marrack, J. R. *The Chemistry of Antigens and Antibodies.* (MRC Special
Report Series No. 230 (1938) and No. 194 (1934).) London: Medical
Research Council.

1939 Tiselius, A. & Kabat, E. A. An electrophoretic study of immune sera and
purified antibody preparations. *J. exp. Med.*, **69**, 119–31.

10

Theories of antibody production

Any theory of antibody production must account for seven facts:
(1) the site of antibody production; (2) the physical nature of
antibodies; (3) the difference between the primary and the
secondary response; (4) immunological memory; (5) the lack of
immunological response to body constituents in normal animals;
(6) immune tolerance; and (7) auto-immune disease.

The history of the theories of antibody production is in
essence the development of successive models each of which
approaches more nearly to the ideal, but is in its turn discarded
in favour of a successor capable of explaining more of the facts.

Our knowledge of the site of antibody formation has
advanced in parallel with the successive theories of antibody
formation and has inevitably influenced the theories. It will be
convenient to consider first the history of beliefs concerning
the site of production of antibodies. The first evidence that
production of antibodies was localized in any special organ was
provided by the work of Pfeiffer and Marx in 1898 when they
showed that the concentration of antibodies was higher in the
spleen and bone marrow of immunized animals than in their
blood. In the following year Deutsch reported experiments with
guinea-pigs in which he found that the development of circulating antibodies was impaired by splenectomy three to five days
after the injection of the antigen. He showed further that the
transplantation of the spleen of an immunized animal into the
peritoneal cavity of a normal animal led to the appearance of
specific antibodies in the blood of the recipient who had never
been injected with the antigen in question.

In 1908 Benjamin and Sluka reported experiments in which
animals had been irradiated with X-rays before and after
immunization. They found that antibody formation was inhi-

bited only when the radiation was administered within a day or two of the dose of antigen. The significance of their results was not clear at the time, but later work was to elucidate its meaning.

In the meanwhile two important developments were made in related fields. In 1911 Landsteiner showed that antibody activity in the serum was associated with the globulin fraction and thus laid the foundation for our knowledge of the chemical nature of antibodies. In 1913 Aschoff and Landau advanced the concept of the reticuloendothelial (RE) system, a group of related cells dispersed throughout the body that was concerned with the phagocytosis of foreign particles, but which was concentrated particularly in the spleen, liver, bone marrow and lymph glands. Aschoff and Landau's concept was to prove particularly fruitful in suggesting lines of experiment designed to determine the site of antibody production.

During the nineteen-twenties a series of elegant experiments making use of the technique of RE blockade focused interest upon the RE system as the source of antibody formation. Animals were injected intravenously with particulate materials, indian ink, colloidal iron or thorotrast, in order to saturate the phagocytic cells of the body, and in the last case to induce radiation damage in these cells selectively, and then various antigens were administered to them and the antibody responses were studied. It soon became clear that this RE blockade impaired the production of antibodies severely, but which cells of the RE system were responsible for the production of antibodies was not revealed by these experiments (Hektoen and Corper, 1920; Bieling and Isaac, 1921; Gay and Clark, 1924; etc.).

Evidence that the cell responsible for the synthesis of humoral antibodies was the lymphocyte came from several different directions. In 1930 Hellman and White reported that there was a great increase in the number of germinal nodes in the spleen and lymph nodes of animals following the injection of antigens. In 1935 McMaster and Hudack showed, by the use of a double antigen technique, injecting one into each ear of a mouse and two hours later amputating the ears, that not only are antibodies present in higher concentration in the lymph nodes draining the site of injection of the antigen than in the blood or other tissues, but further that the antibodies in a lymph node are specific for the antigen injected into its catchment area. In 1938 fresh histological evidence in favour of the lymphocyte as the cell producing the antibodies was adduced by Österland. He showed a correlation between the increase in the number of germinal centres in the lymph glands and the rising titre of antibodies, and also drew attention to the fact that germinal centres were absent in the lymphoid tissues of the foetus and failed to develop in the lymph nodes of germ-free animals.

The year 1939 saw the publication of Tiselius's classic paper in which he demonstrated that antibodies were γ-globulins. While this discovery did not shed immediate light upon the site of antibody synthesis, it made possible the development of techniques which some twenty years later were to pinpoint the origin of antibodies at cellular level.

The Second World War slowed down the rate of progress in pure immunology, but the problems associated with skin grafting the extensive burns suffered by airmen and tank crews formed the starting point for some very significant observations upon the factors influencing the acceptance or rejection of tissue grafts. It was known that autografts (grafts of the host's own tissues) were always accepted and heterografts (grafts of tissue from animals of another species) were always rejected; but homografts (grafts from animals of the same species as the host) would be accepted or rejected according to the genetic relation between the donor and the host, the closer the relationship the greater the probability of acceptance and identical twins always accepting grafts from one another. Studies were made on the process of rejection and it became clear that the histological changes around a graft that was being rejected were very similar to those seen in the lesions of delayed hypersensitivity. While the war lasted no clue as to the fundamental mechanisms that determined acceptance or rejection was found, but in 1945 a lead came from an unexpected direction. In that year Owen published a paper describing cases of 'immunological chimaeras' in dizygotic twin calves. These animals not only possessed red blood cells of incompatible groups, hence the name of the condition, but also were able to accept skin grafts from one another. This was an exception to the previous experience both in man and in experimental animals which indicated that only identical twins were capable of exchanging skin grafts. Owen's observations led Medawar and Billingham to investigate the basis of this specific immune tolerance and in 1953 they published a paper in *Nature* proposing an explanation for Owen's observations and cases of experimental immune tolerance that they themselves had produced in mice from two inbred strains. They postulated a state of immunological maturity which was only attained early in postnatal life. Animals which were immunologically immature reacted to the exhibition of antigen by developing specific tolerance, while immunologically mature animals reacted to the same stimulus by developing specific immunity. In their experiments not only were they able to induce mice from two different inbred strains to exchange skin grafts after prenatal injection of blood cells from one to another, but were able to produce experimental immune tolerance to bacterial antigens, so that the production of antibodies and the development of delayed hypersensitivity failed to take place in animals injected with appropriate antigens during the prenatal or immediately postnatal periods. These studies were important not only for the new knowledge that they produced but for the new problems that they posed. Any future theory of antibody production would have to account in a satisfactory way for the phenomenon of specific immune tolerance. This further constraint was in fact to lead to great theoretical advances.

Between 1953 and 1955 Coons published observations which defined the cells which are the site of antibody production. It was already known that lymph nodes were an important site for the synthesis of specific antibodies; what was not clear was whether the lymphoid cells or the macrophages were the cells involved. Building upon the work of Tiselius, who had shown that antibodies were γ-globulins, this ingenious person developed a technique making use of fluorescence-tagged antibodies against γ-globulin and against the specific antigen used to raise the antibodies. He demonstrated by immunofluorescent microscopy, (*a*) that the cells which were secreting γ-globulin in the lymph nodes of immunized animals were lymphocytes of the type known to histologists as plasma cells and to cytologists as transformed lymphocytes, and (*b*) that the γ-globulin that such cells were secreting was specific antibody that would bind homologous antigen and no other antigens.

Coon's technique has been used by later workers such as White to show that each lymphocyte that is transformed into a plasma cell is probably synthesizing only one specific antibody. In these experiments animals were immunized in the same limb with two distinct antigens, such as killed typhoid bacilli and diphtheria toxoid, and sections of the regional lymph nodes stained with fluorescence-labelled antigens of each type: it was found that individual plasma cells bound either one or the other antigen, but never both.

Just as it appeared that the site of antibody formation was satisfactorily defined, new and complicating evidence came to light. In 1962 Archer and Pierce, and other workers, reported upon the effect of thymectomy on antibody formation in adult and neonatal animals. Their investigations stemmed from the observations of paediatricians that certain cases of congenital agammaglobulinaemia were associated with thymic dysplasia. They found that thymectomy, while without effect upon the subsequent ability to produce specific antibodies in adult animals, was followed by a general immunological paralysis associated with failure to reject homografts in neonatals. It was also shown at about this time that excision of the bursa of Fabricius in birds also impaired immunological responsiveness, but only when done before or immediately after the hatching of the chick from the egg. Bursal excision was later shown to inhibit the production of humoral antibodies, while neonatal thymectomy in mammals affected chiefly the cell-mediated immune mechanisms. The homologue of the bursa of Fabricius in mammals is the gut-associated lymphoid tissue, and it is currently believed that the formation of humoral antibodies is in some way dependent upon this tissue being present at a critical stage in embryonic or neonatal life. It seems likely that this evidence will shed some light upon the mechanisms involved in immunological maturation and the production of specific immune tolerance, but so far the connection has not been made.

The theories of the mechanism of antibody production that have succeeded one another during the last seventy-five years have been determined by the state of knowledge concerning the nature and site of production of antibodies at the time of their promulgation and influenced by the contemporary state of chemical and physical theory.

The first attempt to explain immunity was made by Louis Pasteur in 1880, before antibodies had been thought of. His exhaustion hypothesis regarded the immunity of an animal to a second attack of an infectious disease as due to the exhaustion of an essential nutrient in the tissues of the host. The analogy with the situation in the fermentations, the mechanism of which Pasteur had elucidated so brilliantly, is obvious. Such a theory can explain the specificity of immunity, but is inadequate to explain either immunological memory or the dynamics of the antibody response.

Pasteur's exhaustion hypothesis was abandoned as soon as there was evidence that immunity was connected with the positive phenomenon of antibody production and not with the negative and hypothetical exhaustion of some body constituent. By the middle of the last decade of the nineteenth century enough was known of the properties of antibodies and antigens to enable a theory of immunity based upon the mechanism of antibody production to be advanced.

The first of such theories was the 'side chain theory' of Paul Ehrlich, first advanced in 1897 and elaborated and further elaborated until his death in 1915. Though today this theory is only of historical interest and appears a farrago of dubious biology and undigested chemistry, it was at the time an acceptable model and indeed held the field for some fifteen years after its author's death. The theory depended upon two premises; firstly that protoplasm was composed of a macromolecular backbone and numerous side chains which were the site of specific receptor groups responsible for the assimilation of nutrients and the specific adsorption of dyestuffs, and secondly the principle of compensatory hypertrophy firstly enunciated by Wiegert to account for the reactive overgrowth seen in damaged organs. Ehrlich's model assumed that antigens reacted with their homologous side chains thereby blocking them and rendering them incapable of carrying out their normal role in the economy of the cell; this resulted in a compensatory synthesis of new side chains similar to those inactivated by the toxin, which in turn led, as predicted by the principle of compensatory hypertrophy, to overproduction of the specific side chains, those in excess being sloughed off by the cell into the tissue fluids and bloodstream where they appear as antibodies to the antigen in question. The theory was first propounded with specific toxins such as those of diphtheria and tetanus in mind as typical antigens, and for these it afforded a plausible model. The extension of the theory to take account of newly discovered serological and immunological phenomena over the next fifteen or so years produced a scheme which took account of such things as complement-dependent lysis, the dissociation of toxicity and antigenicity in exotoxins, and natural antibodies. The development was, however, piecemeal and the resulting model was consequently a ramshackle affair, which became less convincing as it became more elaborate. By the time of Ehrlich's death the ramifications of the theory were truly labyrinthine, its terminology esoteric and the very diagrams which were supposed to illuminate the text of its exposition so complex as to be incomprehensible save after prolonged study. Nevertheless it held the field for a further fifteen years and was only superseded when advances in physical and organic chemistry introduced new and more precise concepts into biology in the nineteen-thirties.

Ever since antigen–antibody reactions had been first described they had attracted the interest of certain chemists such as Arrhenius and Madsen. In the early years of this century there were two conflicting theories advanced to explain the observed facts of antigen–antibody reactions. On the one hand there were those who postulated a chemical combination between the two components and used as their model the interaction between weak acids and alkalis: on the other hand there were those who thought of the reactions in purely physical terms and sought an explanation in the properties of colloidal systems. The model chosen implied a different basis for the specificity of antibodies and thus influenced the theories of antibody production.

By the nineteen-thirties enough was known of the primary and secondary structure of proteins for a new theory to be advanced. The idea was that antibodies were globulins synthesized in the presence of antigen molecules in such a way that their secondary folding was determined by the form of the antigen molecule and thus had steric fit with it which accounted for their specificity. This theory of directed synthesis was first advanced by Haurowitz and Mudd and later developed in a more sophisticated form by the great American physical chemist Linus

Pauling in 1940. This came to be known as the direct template hypothesis. The weakness of the theory was that it failed to account for immunological memory in a satisfactory manner. In spite of extensive research there was no evidence for the persistence of antigen in the body for the periods of years which were required to explain the long persistence of this faculty. Work by Pauling and Campbell in 1942 showed that antibody specificity could be abolished by the unfolding of antibody molecules *in vitro*, but nevertheless the theory was manifestly inadequate to explain the phenomena of antibody production as then known.

The next theoretical advance came in 1949 when Burnet and Fenner advanced what is known as the genocopy or indirect template hypothesis. This postulated that after the antibody had been folded to fit the antigen molecule some sort of master template, made probably of RNA, was produced in the cell and thus the synthesis of specific antibody could continue even though all traces of antigen had been eliminated from the system. In the then primitive state of knowledge of protein synthesis this hypothesis appeared a distinct advance upon any previous one, but later when the mechanism of protein synthesis became fully understood it was seen to be inadequate.

Both the theories discussed above were informational, this is to say they assumed that the antigen in some way directs the synthesis of its specific antibody. At the time this seemed the only way to explain how an animal could produce antibodies specific for a wide range of antigens many of which it would never have met in its lifetime.

In 1955 Jerne put forward a hypothesis based upon wholly different assumptions. He abandoned the concept of direct synthesis and claimed to be able to explain the observed facts on the basis of natural selection at the cellular level. He pointed out that the γ-globulins were known to be a heterogeneous population of molecules and postulated that in the bloodstream small quantities of these proteins capable of reacting specifically with a very wide range of antigenic groups were present. On the injection of an antigen it would combine with its homologous γ-globulin and the ingestion of this antigen–antibody complex by immunologically competent cells would lead, in some mysterious way to the synthesis of further specific antibody. Jerne's hypothesis can now be seen to be a sort of halfway house between

Fig. 10.1. F. Macfarlane Burnet (1899—).
From C. E. Lyght, *Reflections on Research and
the Future of Medicine*, p. 9. New York,
McGraw-Hill (1967).

the purely instructional hypothesis that proceeded it and the
hypothesis based solely upon mutation and selection that
followed it.

It was Burnet (Fig. 10.1) who developed Jerne's ideas to their
logical conclusion. In 1957 he published his clonal selection
hypothesis in which he attempted to account for all the known
phenomena of immunity, including specific immune tolerance,
invoking only the mechanisms of mutation and selection at
cellular level. The way in which mutation and selection could
account for such improbable facts as the development of
resistance to newly discovered anti-microbial drugs in bacterial
populations' exposed to their action for the first time, had
impressed all microbiologists with the potentialities of such a
model. Burnet postulated a very large, but not infinite, number
of types of immunologically competent cells developing during
the later stages of foetal life and their diversity being further
recruited by continuing somatic mutation through life. Contact
with antigen that was homologous with the type of γ-globulin

that a cell was genetically programmed to produce would result, according to the dose, in either cell division or cell death. Cell division would give rise to a clone which was composed of cells all of which would synthesize the specific γ-globulin homologous with the antigen, that is specific antibody. This model can account for the difference between the primary and the secondary response, for long-term immunological memory, for specific immune tolerance and the recognition of self on the assumption that a dose of antigen which causes cell division in the adult animal will be lethal when brought into contact with immature immunocytes. The production of specific immune paralysis by the administration of very large doses of antigen to adult animals fits in well with the theory. The phenomena of auto-immune disease can be accounted for either by assuming that certain body components, such as the aqueous humour of the eye or thyroglobulin, are not normally accessible to the immature immunocytes of the foetus, and so fail to destroy the cells potentially capable of giving rise to clones secreting antibody specific to them; or alternatively by assuming that such forbidden clones arise in adult life by somatic mutation. The clonal selection hypothesis of Burnet is the most satisfactory model for the synthesis of antibodies that has been advanced to date. There are, however, a number of phenomena such as the passive transfer of delayed hypersensitivity to tuberculin by cell-free material and the failure to induce a primary response to any antigen in cultures of lymphocytes, which are not satisfactorily accounted for: as a consequence several theories have been advanced which, while more satisfactory in explaining certain special cases have not yet superseded it, as they are either developments of the clonal selection theory, or lack its general power.

The two more recent theories that are perhaps of greatest interest are Lederberg's subcellular selection hypothesis and Fishman's transformation hypothesis. Lederberg assumes that every immunocyte has the potentiality to synthesize a great variety of globulins of differing specificity. The fact that antibody specificity is now known to be dependent upon the amino acid sequence at the N end of the Fab part of the molecule and that the DNA region of the genome which codes for this might well be a mutational hot spot, such as are known to occur in certain regions of the bacterial and phage genomes, gives some probability to this assumption. Lederberg's theory further assumes that the entry of antigen into a cell which was synthesizing the homologous γ-globulin would lower the intracellular concentration of the antibody as antigen–antibody complex was formed and would thus, assuming that the mechanism responsible for the synthesis of the component polypeptide chains of the antibody was, in accordance with the law of mass action, stimulated to produce more, cause an overshoot of the normal level and account for the production of a sufficient excess to produce circulating antibody. This theory is clearly derived from the models of Monod and others which have been so successful in explaining the phenomena of adaptive enzyme production in bacteria. In Lederberg's view the antigen is acting as an 'inducer' for the synthesis of the specific antibody. This theory has the advantage of integrating the theories of adaptive enzyme formation and antibody formation, but it is difficult to see how it accounts for immune tolerance, self-recognition and auto-immune disease. Fishman has shown experimentally that antibody production in response to a primary stimulus can be achieved *in vitro* provided that the antigen is first ingested by macrophages and these activated macrophages are then added to a suspension of lymphocytes. He has further shown that RNA passes from the activated macrophages to the lymphocytes before the latter can begin to synthesize antibody. He has therefore put forward what is a semi-instructional theory backed with experimental evidence. It is curious that very little notice has been taken of Fishman's results by other writers on theoretical immunology.

A full and perfect theory of antibody production must await further experimental evidence and in particular more evidence about the phylogeny of the immune response and the different types of antibodies. The work done to date on the evolution of antibody formation has shown that humoral antibodies are confined to vertebrates though cellular immunity and therefore possibly cell-bound antibodies have been found in all phyla examined. Much more detailed results are needed before this type of evidence can shed more light upon the the problem of the mechanism of antibody production.

REFERENCES

1880 Pasteur, L. Sur les maladies virulentes, et en particulier sur la maladie appelée vulgainment choléra des poules. *C. r. Séanc. Acad. Sci.*, **90**, 239–48.

1899 Deutsch, L. Contribution a l'étude de l'origine des anticorps typhiques. *Ann. Inst. Pasteur*, **13**, 689–727.

1898 Pfeiffer, R. & Marx, K. F. H. Die Bildungsstätte der Choleraschutzstoffs. *Z. Hyg. InfektKrankh.*, **37**, 272–97.

1900 Ehrlich, P. Croonian lecture – 'On immunity with special reference to cell life'. *Proc. R. Soc. Ser. B.*, **66**, 424–8.

1900 Landsteiner, K. Zur Kenntniss der antifermentativen, lytischen und agglutinierenden Wirkungen des Blutserums und der Lymphe. *Zentralbl.Bakt.*, **27**, 357–62.

1902 Arrhenius, S. *Physical chemistry Applied to Toxins and Antitoxins.* (Festschrift Staatens Serum Inst. NO. 3.) State Serum Institute.

1904 Arrhenius, S. & Madsen, T. Toxines et antitoxines. Le poison diphtérique. *Zentralbl.Bakt.*, **36**, 612–24.

1908 Benjamin, E. & Sluka, E. Antikörperbildung nach experimenteller Schädigung des hämatopoetischen Systems durch Röntgentstrahlen. *Wien. klin. Wochenschr.*, **21**, 311–13.

1911 Landsteiner, K. & Praisek, E. Uber die Beziehung der Antikörper zu der präzipitablen Substanz des Serums. *Z. ImmunForsch. exp. Ther., Orig.*, **9**, 68–102.

1913 Aschoff, L. & Landau. In *Lectures on Pathology*, by L. Aschoff, p. 2. New York: Paul Hoeber (1924).

1920 Hektoen, L. & Corper, H. J. The influence of Thorium X on antibody formation. *J. infect. Dis.*, **26**, 330–5.

1922 Bieling, R. & Isaac, S. Experimentelle Untersuchungen über intravitale Hämolyse. IV. Die Bedeutung des Reticulo-Endothels. *Z. exp. Med.*, **28**, 180–92.

1922 Hektoen, L. & Corper, H. J. Effect of injection of active deposit of radium emanation on rabbits with special reference to the leucocytes and antibody formation. *J. infect. Dis.*, **31**, 305–12.

1924 Gay, F. P. & Clark, A. R. The reticulo-endothelial system in relation to antibody formation. *J. Am. med. Ass.*, **83**, 1296–7.

1930 Hellman, T. & White, G. Das Verhatten des lymphatischen Gewebes während eines Immunisierungsprozesses. *Virchows. Arch. path. Anat. Physiol.*, **278**, 221–57.

1932 Mudd, S. A hypothetical mechanism of antibody formation. *J. Immun.*, **23**, 423–7.

1935 McMaster, P. D. & Hudack, S. S. The formation of agglutins within lymph nodes. *J. exp. Med.*, **61**, 783–805.

1938 Österland, G. Die Reaktion des lymphatischen Gewebes während der Ausbildung der Immunität gegen Diptherietoxin. *Acta path. microbiol. Scand.*, **34**, 1–139.

1939 Tiselius, A. & Kabat, E. A. An electrophoretic study of immune sera and purified antibody preparations. *J. exp. Med.*, **69**, 119–31.

1940 Pauling, L. A theory of the structure and process of formation of antibodies. *J. Am. chem. Soc.*, **62**, 2643–57.

1941 Burnet, F. M., Freeman, M., Jackson, A. Y. & Lush, D. *The Production of Antibodies.* (Monographs of the Walton and Eliza Hall Institute No. 1.) London: Macmillan & Co. Ltd.

1945 Owen, R. D. Immunogenetic consequences of vascular anastomoses between bovine twins. *Science*, **102**, 400–1.

1949 Burnet, F. M. & Fenner, F. *The Production of Antibodies.* London: Macmillan & Co. Ltd.

1953 Coons, A. H., Leduc, E. H. & Conolly, J. M. Immunohistochemical studies of antibody response in the rabbit. *Fedn. Proc. Fedn. Am. Socs. exp. Biol.*, **12**, 439.

1953 Billingham, R. E., Brent, L. & Medawar, P. B. 'Actively acquired tolerance' of foreign cells. *Nature, Lond.*, **172**, 603–6.

1955 Coons, A. H., Leduc, E. H. & Connolly, J. M. Studies on antibody production. I. A method for the histochemical demonstration of specific antibody and its application to a study of the hyperimmune rabbit. *J. exp. Med.*, **102**, 49–59.

1955 Jerne, N. K. The natural selection theory of antibody formation. *Proc. Acad. Sci., Paris*, **41**, 849–57.

1959 Burnet, F. M. *The Clonal Selection Theory of Acquired Immunity.* London: Cambridge University Press.

1959 Lederberg, J. Genes and antibodies. *Science*, **129**, 1649–53.

1959 Fishman, M. Antibody formation in tissue culture. *Nature, Lond.*, **183**, 1200–1.

1961 Fishman, M. Antibody formation *in vitro. J. exp. Med.*, **114**, 837–56.

1962 Archer, O. K., Pierce, J. C., Papermaster, B. W. & Good, R. A. Reduced antibody response in thymectomized rabbits. *Nature, Lond.*, **195**, 191–3.

1963 Fishman, M., Hammerstrom, R. A. & Bond, V. P. *In vitro* transfer of macrophage RNA to lymph node cells. *Nature, Lond.*, **198**, 549–51.

11

The classification of bacteria

The classification of bacteria must be considered as a part of the general taxonomy of living things. We use the same terms as botanists and zoologists and the class Schizomycetes which comprehends all the bacteria is a part of the greater hierarchy of biological classification. It is therefore essential to consider some of the general characteristics of biological classifications if we are to understand how the science of bacterial taxonomy has developed.

Classifications, as opposed to catalogues or keys, may be natural or artificial. From the earliest days of biological classification scientists have attempted to devise natural classifications in preference to artificial ones. In bacteriology this has led to attempts to show that what were in reality convenient artificial classifications should be regarded as natural. The first attempt at a comprehensive classification of living things, that of Linnaeus, the *Systema Naturae* claimed to be a natural classification in the sense that it revealed the master plan which guided the Almighty during the process of the creation of the world. Linnaeus wrote 'It is...the business of a thinking being, to look forward to the purposes of all things; and to remember that the end of creation is, that God may be glorified in all his works.' Linnaeus believed in the fixity of species following the original and definitive creation of the world in six days: it was therefore not unreasonable, even if today it appears somewhat blasphemous, for him to attempt to look into the mind of God and deduce the plan upon which he had worked.

When the account of the creation given in the book of Genesis was discredited and the scientific world adopted Darwin and Wallace's evolutionary account of the origin of species, the metaphysical basis of a natural classification was changed. Under the new dispensation the term meant a classification that reflected the phylogeny of the organisms being classified; those species recently descended from a common ancestor being grouped together in a genus, and those whose common ancestor was further back being placed in different genera, families or orders according to the distance that separated them from the common ancestor.

An important feature of Linnaeus' classification was its weighting of the characters used to identify the organisms. This is seen most clearly in his botanical classification which depends upon the character of the flowers, the sexual apparatus of the plants, almost to the exclusion of all other characters. While the use of weighted characters is clearly convenient if one wishes to evolve a dichotomous system based upon formal Aristotelian logic, there are obvious objections from a biological point of view to placing differential weight upon one type of character as opposed to another. It was not long before an attempt was made to produce a system of classification based upon the use of all characters, each being given equal weight. In 1763 Michael Adanson published his *Familles des Plantes* which was based upon such an unweighted system. The species, genera and families of plants which he proposed were not very different from those put forward by Linnaeus, but the basis for their recognition was quite different. Adanson's classification did not become popular, partly because it was more difficult to use than that of Linnaeus, and partly because of the immense prestige of the Swedish botanist. Adanson's work remained little-known until very recently when, with the development of computer programmes for bacterial taxonomy his work was rediscovered. An un-weighted system of classification is much more suited to computerization than a weighted one and his principles were adopted by Sneath and others as the basis of the new computerized or Adansonian taxonomy.

In Linnaeus' classification the infusoria, which included the animalcules described by van Leeuwenhoek, some of which we believe to have been bacteria, were placed in the genus *Chaos*, and there are cynics who would say that this was a more honest description of the state of knowledge than the compendious and conflicting systems of bacterial classification which abound today.

Fig. 11.1. C. G. Ehrenberg (1795–1876).
From an unpublished, undated line drawing
by J. P. Singer in the Wellcome Institute. By
courtesy of 'The Wellcome Trustees'.

The next step came when O. F. Muller published his *Animalcule Infusoria et Marina* in 1786. He proposed a division into two
genera, *Monas* and *Vibrio*. The latter has stood the test of time,
the former not.

In 1838 the great German biologist and traveller C. G.
Ehrenberg (Fig. 11.1) published a monograph entitled *Die
Infusionstierchen als vollkommene Organismen*. In this work he
proposed a classification of the infusoria in which they were
divided into two families, Monas and Vibriona. The family
Vibriona was further divided into five genera: *Bacterium*, *Vibrio*,
Spirochaeta, *Spirillum* and *Spirodiscus*. Each of these genera
contained one or more species. One of Ehrenberg's species is of
great historical interest. The type species of the genus *Bacterium*
was an organism that he named *Bacterium triloculare*. This
organism he had observed in the waters of the oasis of Aswan in
Egypt, and he gave detailed description of its morphology. For

more than one hundred years it remained the type species of the genus *Bacterium*. Unfortunately in those days it was not possible to obtain pure cultures of bacteria and therefore it does not occur in any of the world's collections of type cultures: indeed it has never been seen again since Ehrenberg observed it. Finally, only a few years ago the taxonomic scandal of defining a genus in terms of a non-existent type species became so clamorous that the editors of *Bergey's Manual of Determinative Bacteriology* expunged the species from the record. It is perhaps only in bacterial taxonomy that such an anomaly would have been tolerated for so long.

All the taxonomists that had dealt with the bacteria up to this time had regarded them as belonging to the animal kingdom. Van Leeuwenhoek had coined the term 'animalcule' for the organisms that he had observed with his microscopes and the assumption that these minute things were animal infusoria had persisted. The first person to appreciate the true nature of the bacteria was the German botanist Nageli. In an article published in the *Botanischer Zeitung* in 1857 he proposed that they should be regarded as a class on their own within the vegetable kingdom for which he coined the name Schizomycetes or fission fungi.

Between 1872 and 1876 Ferdinand Cohn, Professor of Botany at the University of Breslau, published a new system of classification for the bacteria. This was based, as were all previous classifications, upon morphological criteria alone. Cohn was a brilliant microscopist and his published drawings of bacteria are of unsurpassed elegance and clarity. He divided the bacteria into four tribes each of which was made up of one or more genera:

Tribe I Sphaerobacteria	Genus I *Micrococcus*
Tribe II Microbacteria	Genus II *Bacterium*
Tribe III Desmobacteria	Genus III *Bacillus*
	Genus IV *Vibrio*
Tribe IV Spirobacteria	Genus V *Spirillum*
	Genus VI *Spirochaeta*

The genus *Bacillus* contained two species, *Bacillus subtilis* and *Bacillus anthracis*. Cohn was the first person to describe bacterial endospores and to note their heat-resistance.

The methodological revolution ushered in by Robert Koch's

development of solid media and the consequent possibility of obtaining pure cultures for study had profound effects upon bacterial taxonomy: no longer was it only morphological characters which formed the basis of classifications, cultural and biochemical results became the basis of criteria just as, or more important, than these. By the last decade of the nineteenth century an immense amount of data concerning cultural, biochemical and pathogenic behaviour of various bacteria had been assembled. Even more important, the objectives of classification had altered: the older systematists had been concerned only to reduce the multiplicity of life forms to some order, whereas the new bacteriology was concerned with the specific problems of recognizing the organisms responsible for infectious diseases such as typhoid, diphtheria or tuberculosis, and differentiating these dangerous organisms from the similar but harmless saprophytes. Nevertheless the main criteria for the identification and classification of bacteria remained morphological until early in the twentieth century.

Two important new classifications appeared during the last decade of the nineteenth century: that of Migula, first published in 1894 and revised in 1900, and that of Lehmann and Neumann published in 1896. Migula divided the Schizomycetes into two orders; the Eubacteria, 'non-nucleated cells colourless or coloured slightly, without sulphur granules or bacteriopurpurin, without chlorophyll'; and the Thiobacteria, 'non-nucleated cells containing sulphur granules or bacteriopurpurin'. The Eubacteria were subdivided on morphological grounds into three families; the Coccaceae, Bacteriaceae and Spirillaceae. Of the medically important genera he recognized only *Streptococcus*, *Bacterium* (defined as non-motile rods), *Bacillus* (defined as rods motile by means of peritrichous flagella), *Pseudomonas*, *Spirillum* and *Spirochaeta*. Migula attempted to produce a logically consistent general classification of all the known types of bacteria based entirely upon morphological characters, ignoring the new methods of differential staining developed by Neelsen and Gram and taking no account of such important characters as spore formation.

Lehmann and Neumann in their *Atlas und Grundriss der Bakteriologie* adopted a more empirical approach and considered the staining reactions and the ability to form endospores

Fig. 11.2. C. E. A. Winslow (1877–1957).
From *J. Bact.* (1957), **73**, facing p. 295.

as formal diagnostic features in their system. They proposed
to divide the Schizomycetes into two orders; the Schizomyce-
tales containing three families of unbranched organisms, the
Coccaceae, the Bacteriaceae and the Spirillaceae; and the
Actinomycetales containing two families of organisms showing
true branching, the Proactinomycetaceae (*Corynebacterium* and
Mycobacterium) and Actinomycetaceae. Within the family Bacteriaceae there were two genera; *Bacterium*, without endospores,
and *Bacillus* with endospores. The mycobacteria were differentiated from the corynebacteria on the grounds of their acid-fast
staining.

It is curious that it was not until 1908 that pathogenicity was
admitted as a taxonomic criterion. In that year the Winslows
(Fig. 11.2), in a monograph on the systematic relations of the
Coccaceae, suggested that they be separated into two sub-
families; the Paracoccaceae which included all parasitic genera

and the Metacoccaceae which included the saprophytic genera. They also admitted as differential characters between genera the reaction with Gram's stain and the fermentation of various carbohydrates. Once again we have an empirical classification designed to facilitate the differentiation between medically and industrially important species and others.

Orla Jensen in 1909 published a system of classification which while admitting morphological and pathogenic characters was based primarily upon the metabolic activities of the bacteria. He divided the Schizomycetes into two orders; the Cephalotrichinae whose 'life energy was derived mainly from oxidative processes without producing notable amounts of unoxidized split products' and the Peritrichinae in whose case 'splitting of carbohydrates or amino acids was the primary role in metabolism rather than oxidation or denitrification'. This effort to produce a rational biochemical or metabolic classification was premature and most of the families that he suggested have lapsed into oblivion; all the same his effort had considerable influence upon later taxonomists and must stand as an important step forward in the classification of bacteria.

From 1915 onwards the American bacteriologist Buchanan published a series of papers, mainly in the *Journal of Bacteriology*, under the general title of 'Studies on the nomenclature and classification of the bacteria'. In these papers he arranged the bacteria into families, tribes and genera, making use of a wide range of characters – morphological, tinctorial, biochemical, and pathogenic. The influence of Buchanan's work was far-reaching. In 1917 the Society of American Bacteriologists set up a Committee on Characterization and Classification of Bacteria. Its two reports of 1917 and 1920 led directly to the compilation of the most famous of modern books on bacteriological classification, *Bergey's Manual of Determinative Bacteriology*. The editors defined their object in the introduction to the first edition as 'to make the system of classification promulgated by the Society of American Bacteriologists of greater value to students by extending the classification to the individual species of the genera that have been recognized as valid by the Committee'. The manual is an extensive key for the recognition of strains of bacteria down to the specific level, but it has proved even more important as a taxonomic system. Seven editions of this work have been published and it has undergone progressive development: in the first edition there were only five orders recognized but by the seventh edition ten orders appear. The elaboration of the system extended also to genera and species. The manual remains today an essential tool for the working bacteriologist in spite of the fact that many students no longer accept the type of weighted classification that underlies it and other keys have been published some of which, such as Cowan and Steel's *Manual for the Identification of Medical Bacteria* are widely used.

During the third and fourth decades of the twentieth century a number of alternative classifications were published, the most important of which was that of Kluyver and Van Niel (1936). In view of the interest of the Delft school in comparative biochemistry, it is not surprising that this system gave considerable weight to the nutritional and metabolic characters of the micro-organisms. They recognize four families; the Pseudomonadaceae, the Micrococcaceae, the Mycobacteriaceae, and the Bacteriaceae. These are subdivided into genera in accordance with the fundamental metabolism of the species; photoautographic, photoheterotrophic or chemoautotraophic. The system is only published as an outline and does not go further than generic level. An interesting application of the nutritional metabolic approach to classification taken as far as the species level is the monograph *Bacterial Nutrition. Material for a Comparative Physiology of Bacteria* published by B. C. J. G. Knight in the same year.

Two developments in other fields of science have altered our outlook on bacterial classification during the last decade. The development of the electron microscope to the point where it is posible to examine ultra-thin sections of individual cells has led to an appreciation of the fundamental differences between the organization of the bacteria and the blue-green algae on the one hand and the cells of the higher plants, animals and fungi on the other. The differences were brought sharply into focus when in 1957 E. C. Dougherty of Stanford proposed the terms procaryotic and eucaryotic to describe the type of cellular organization of the bacteria and blue-green algae, and the higher cells respectively. It is now clear that this division is of great evolutionary significance and almost certainly pre-dates the emergence of the first animals and green plants. This new knowledge has not of course led to any internal rearrangement within the class Schizomycetes, but it has defined the place of the class in the natural system as a whole in a way never before possible.

The rapid advance in the design of high-speed electronic computers was the other development that altered the face of bacterial taxonomy. The application of these instruments to bacterial classification has been pioneered by Sneath. He realized the power of computers to handle masses of data which would previously have been impossible to deal with in the lifetime of any one worker. He appreciated the need for an unweighted classification to fit with the 'off/on' switching of the machines and he proceeded to develop such a system based upon the long-neglected work of Michael Adanson. Sneath and others influenced by his work (Hill; Barnes and Goldberg; Davies and Newton) have applied unweighted Adansonian classifications to a number of different families and genera. The computer programmes that they have devised arrange the different species in order of similarity, and their relationship is quantified in terms of a 'similarity coefficient' which takes values between 0 and 1 according to how many characters the two species have in common. This work has led to the emergence of the concept of the 'taxon', a group of life organisms with a high similarity coefficient, separated from other taxons by a low similarity coefficient. The taxon is an operational entity and makes no assumptions about the natural relations between groups of organisms, and to that extent it is an advance upon older 'natural classifications', which in the case of the bacteria are bound to fail to conform to the definitions in use in general biology. It is, however, rather disappointing that the taxons generated by the new computer programmes are almost always similar to the genera and species of the older classifications. This result is hardly surprising when one remembers that the characters used to identify the organisms are the same in each case.

There is one field where new and different characters are available for the identification of strains of bacteria; that of bacterial genetics. Here it is possible to classify the micro-organisms in terms of mating distance. Such an approach yields a very different picture from the classical or neo-Adansonian classifications based alike upon phenotypic characters alone. It is found that fertile crosses can be achieved not only between bacteria allotted to different species but between organisms such as vibrios and salmonellae which are placed classically in different orders. This plainly makes nonsense of the terms species, genus, family or order, in their usual sense: on the other hand the established groups are of undoubted pragmatic value in the various fields of applied bacteriology that they serve. It now seems possible that a truly natural classification could be developed on the basis of an extended genetic study, but such a system would be of little use to the medical or dairy bacteriologist who must continue to differentiate between useful and harmful organisms in the context of the control of infectious disease or the production of cheese and other milk products.

In the future we can at least begin to accept the fact that the practical working classifications of bacteria must be artificial and empirical and that it is therefore futile to hope to synthesize some general natural system from such elements. It may be that over the next generation enough genetic information will accumulate to enable a natural general classification to be developed, but even should this happen, practising bacteriologists will have to go on using the limited pragmatic classifications that have been developed to deal with the problems of the different branches of applied bacteriology.

REFERENCES

1743 Linnaeus. *Genera Plantarum.* Paris.

1763 Adanson, M. *Familles des Plantes.* Paris.

1786 Muller, O. F. *Animalcula infusioria et marina systematice descripsit et ad vivum delineari curavit O.F.M. opus cura O. Fabricii, 4°,* Hauniae.

1838 Ehrenberg, C. G. *Die infusionsthierchen als vollkommene Organismen.* Leipzig.

1872 Cohn, F. Untersuchungen über Bakterien: I. *Beitr. Biol. Pfl.,* **1,** Heft 3, 127–222.

1875 Cohn, F. Untersuchungen über Bakterien: II. *Beitr. Biol. Pfl.,* **1,** Heft 3, 141–207.

1877 Koch, R. Untersuchungen über Bakterien: V. Die Aetiologie der Milzbrand Krankheit, begrandet auf Entwiklungsgeschichte des *Bacillus anthracis. Beitr. Biol. Pfl.,* **2,** 277–308.

1894 Migula, W. Ueber ein neues System der Bakterien. *Arb. Bakt. Inst. tech. Hochsch. Karlsruhe,* **1,** 235–8.

1896 Lehmann, K. B. & Neumann, R. O. *Atlas und Grundriss der Bakteriologie.*

1906 Winslow, C. E. A. & Rogers, A. F. A statistical study of gnetic characters in the Coccaceae. *J. inf. Dis.*, **3**, 485–546.

1909 Jensen, O. Die Hauptlinien des naturlichen Bakteriensystems. *Zentralbl. Bakt.*, **22**, 305–46.

1915 Buchanan, R. E. Nomenclature of the Coccaceae. *J. inf. Dis.*, **17**, 528–41.

1923 *Bergey's Manual of Determinative Bacteriology.* Baltimore: Williams & Wilkins Co.

1936 Kluyver, A. J. & Van Niel, C. B. Prospects for a natural system of classification of bacteria. *Zentralbl.Bakt.* (Abt. II) **94**, 369–403.

1936 Knight, B. C. J. G. *Bacterial Nutrition. Material for a Comparative Physiology of Bacteria.* (MRC Special Report Series No. 210.) London: HMSO.

1948 *Bergey's Manual of Determinative Bacteriology*, 6th edition. London: Baillière, Tindall & Cox.

1957 *Bergey's Manual of Determinative Bacteriology*, 7th edition. London: Baillière, Tindall & Cox.

1957 Dougherty, E. C. Neologisms needed for structure of primitive organisms. I. Types of nuclei. *J. Protozool.*, **4** (Supplement), 14.

1957 Sneath, P. H. A. The application of computers to taxonomy. *J. gen. Microbiol.*, **17**, 201–26.

1959 Hill, L. H. The Adansonian classification of the staphylococci. *J. gen. Microbiol.*, **20**, 277–83.

1965 Cowan, S. T. & Steel, K. J. *Manual for the Identification of Medical Bacteria.* London: Cambridge University Press.

1968 Barnes, E. M. & Goldberg, H. S. The relationship of bacteria within the family Bacteroidaceae as shown by numerical taxonomy. *J. gen. Microbiol.*, **51**, 312–24.

1969 Davies, G. H. G. & Newton, K. G. Numerical taxonomy of some named coryneform bacteria. *J. gen. Microbiol.*, **56**, 195–214.

1969 Tsukamara, M. Numerical taxonomy of the genus *Nocardia. J. gen. Microbiol.*, **56**, 265–87.

12
Virology

The technique of purification of mixed cultures of bacteria by plating out on solid media developed by Robert Koch ushered in an era of rapid progress in the field of medical bacteriology. By 1890 the causative organisms of the majority of the common infectious diseases of men and animals had been isolated and studied, and rational preventive measures were being developed. There remained however a number of common epidemic diseases – measles, smallpox, chickenpox and influenza, to name but a few – where it had not proved possible to isolate any organisms that satisfied the postulates laid down by Koch and could therefore be assigned a causative role. In addition to the human diseases there were many animal and plant diseases where no bacterial pathogen had been implicated.

The development of reliable bacteriological filters by Chamberland had led to the discovery of the exotoxin of the diphtheria bacillus, and the tetanus bacillus, so that it was natural that in cases where it had proved impossible to isolate bacteria from the lesions of a communicable disease, a search for toxins should be made by injecting cell-free filtrates of infected tissues into susceptible hosts. It was a result of such work that the first cases of diseases caused by viruses were discovered.

The world virus has been in use for some hundreds of years. Originally the Latin word for a poison, it had come by the sixteenth century to be used for the venom of poisonous beasts and by the eighteenth century was used to mean 'a poisonous substance in the body as a result of some diseases, especially one capable of being introduced into other persons or animals by inoculation or otherwise and of developing the same disease in them (1728)'. By 1800 we find reference to 'the smallpox virus' present in the pustules. Louis Pasteur in the eighteen-sixties

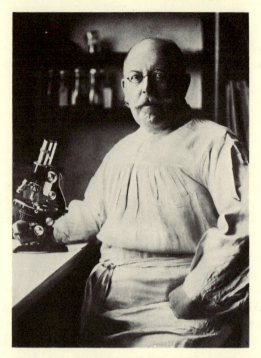

Fig. 12.1. Friedrich Loeffler (1852–1915). From an original photograph in the Wellcome Institute, presented by Dr Lieven in 1927. By courtesy of 'The Wellcome Trustees'.

used the term virus to refer to any living agent causing infectious disease; he spoke for instance of the virus of fowl cholera, as well as the virus of rabies. The discovery that there were infectious diseases of plants and animals which could be transmitted by inoculation of the cell-free filtrates from the lesions, led to the coining of the term filterable virus. Only later was the qualifying adjective dropped and the word virus alone used to refer to these entities.

The first disease to be shown to be caused by a filterable virus was tobacco mosaic disease. Iwanowski published a paper in 1892 in which he reported that this condition could be passed from plant to plant in series by the inunction into leaves of healthy plants of cell-free filtrates from the leaves of plants showing the characteristic lesions of the disease.

In 1898 Loeffler (Fig. 12.1) and Frosch reported the first

example of an animal disease where a filterable virus appeared to be the causative organism: foot-and-mouth disease. Here once again the evidence was the serial transmission of the disease from animal to animal by means of the inoculation of cell-free filtrates from the lesions, such filtrates being sterile and showing no organized material on microscopy; hence the alternative designation of the new agents as ultra-microscopic viruses.

In the same year S. M. Copeman published a monograph on vaccination in which we find one of the earliest references to the use of the fertile hen's egg as a method for the cultivation of viruses. It is interesting to speculate as to why this technique was not further developed at this time. Possibly the fact that Copeman himself did not see the general applicability of the method may have had something to do with this and the lack of knowledge concerning other virus diseases must also have been a factor. There was still a tendency in cases where the evidence for a viral aetiology was entirely negative to attribute this to technical failure and to hope for the isolation of some hypothetical bacterial pathogen in the near future.

At the beginning of the new century the heroic investigations of the American Yellow Fever Commission led by Walter Reed and the later work of the French Commission proved the viral aetiology of yellow fever, the first of the arboviruses to be recognized. The insect transmission was suggested by the analogy with malaria, while the viral nature of the infectious agent was proved by the production of the disease in human volunteers by the injection of a cell-free filtrate of the blood of a patient suffering from the disease.

The fertile egg was used once again for the cultivation of a virus by Payton Rous in his work on the fowl sarcoma that bears his name. Although the tumour was proved to be caused by a virus the technique still remained limited to the special field of virus-induced tumours of poultry and no one seems to have considered applying it for the cultivation of other viruses.

By 1911 the point had been reached where, using the criterion of filterability of the infecting agent alone, plant and animal viruses had been recognized, the fact that certain viruses could be transmitted by insect vectors had been established, and at least one oncogenic virus had been identified. The cultivation of viruses was possible only in intact animals except for the two

Fig. 12.2. F. W. Twort (1877–1950).

special cases where the fertile hen's egg had been brought into use. The recognition of a virus depended upon the clinical signs in the infected animal, characteristic macroscopic lesions in experimental animals at post-mortem and in certain cases, notably rabies and smallpox, upon the microscopic detection of inclusion bodies in the cells of the affected tissues. It is not surprising that many diseases which were thought to be viral on epidemiological or negative bacteriological grounds remained listed as 'of uncertain aetiology'.

In 1915 a new class of viruses affecting neither plant nor animal cells but bacteria was discovered. F. W. Twort (Fig. 12.2) reported in *The Lancet* his observations upon certain staphylococcal cultures. He noted that their colonies on agar underwent localized degenerative changes. He showed that it was possible to transmit this change from one culture of the susceptible organism to a fresh one by dropping into the surface of the fresh culture a single drop of much diluted cell-free filtrate of the first culture. Serial transmission could be continued for as long as required. Twort did not realize that he was dealing with a viral disease of bacteria and his speculations as to the cause of the phenomenon that he had observed did not contribute to our understanding of the cause of the transmissible degenerative change in the colonies.

In 1917 D'Herelle published an account of another curious phenomenon – the lysis of cultures of dysentery bacilli growing in broth following the addition of a few drops of a cell-free filtrate prepared from a mixed culture made by inoculating the faeces of patients suffering from bacillary dysentery into broth. He was able to transmit the lysis from one culture of dysentery bacillus to another by addition of cell-free filtrates from a lysed culture. Over the next thirteen years D'Herelle published a series of papers on this phenomenon and came to the conclusion that it was caused by a filterable virus parasitic upon the bacteria. It was he who coined the term bacteriophage. At first he thought there was but one bacteriophage, but it later became clear that there must be a group of bacterial viruses differing in their host specificity. Throughout the nineteen-twenties and early nineteen-thirties there was controversy as to the nature of the bacteriophage. Not all bacteriologists were convinced that virus was involved and the evidence available at that time was not sufficient to settle the matter unequivocally.

The next advance in virology came as a result of the development of the ultra-violet microscope by the English bacteriologist Barnard. This instrument, making use of quartz lenses and photographic recording doubled the resolving power of the traditional light microscope and made possible important studies upon the nature of the inclusion bodies that were recognized to be signs of the presence of virus in cells. It became clear that the inclusion bodies were aggregates of smaller elementary particles surrounded by host material. The elementary bodies were taken to be the actual virus particles but their further structure could not be investigated as they lay at the limit of the resolving power of the ultra-violet microscope.

The development of tissue culture techniques in the nineteen-twenties led to their application for the cultivation of viruses. In 1925 Parker and Nye succeeded in growing *Herpes simplex* virus and *Vaccinia* virus in explant cultures of rabbits testicles, and two years later Carrell and Rivers reported successful results with cultures of minced chick embryo tissue. These early experiments called for great technical ingenuity and a scrupulous aseptic technique, and it reflects great credit upon the workers that they were able to achieve undoubted multiplication of viruses in *in vitro* systems with the methods then available. There are two points that are of particular significance that should be noted. In the first place, these early workers believed that viruses would only multiply in actively dividing cells. In the second place they were without the benefit of antibiotics to control bacterial contamination of their cultures and were thus severely limited as to the type of specimen that they could attempt to culture virus from.

The work of the Maitlands (Fig. 12.3), published in 1928, constituted a major advance in the field of tissue culture cultivation of viruses. They showed that it was possible to obtain multiplication of virus in suspensions of finely chopped hen's kidney, thus proving that viruses would grow in non-multiplying cells. That their technique did not lead immediately to further developments was due to two factors; the necessity for tedious indirect methods for measuring viral growth, and the lack of available antibiotics to control contamination of the cultures by bacteria. It remained a research method and could not be applied to diagnostic virology or to the preparation of vaccines until antibiotics became available in the late nineteen-forties. It is an interesting speculation as to why Maitland did not take up and further develop his method at that time.

At last in 1931 the potentialities of the fertile egg for the culture of viruses were appreciated. The publication of a paper on the use of the fertile hen's egg in virology, by Goodpasture in America, sparked off a spate of papers dealing with the cultivation of various viruses in the chick embryo. It was, however, the break-through achieved in the case of influenza virus which dominated the scene. Much of the credit must go to the Australian Macfarlane Burnet, who applied Goodpasture's techniques successfully to the growth of influenza virus within a few months of the publication of his article. Up to that time the only way to cultivate the virus had been by infecting ferrets, and antibodies could only be detected by protection tests using the same vicious animals. The value of the hen's egg as a culture

Fig. 12.3. H. B. Maitland (1895–1972). From *J. Med. Microbiol.* (1953), **6**, facing p. 254.

medium was greatly enhanced by the discovery of viral haemagglutination by Hirst in 1941. With pox viruses it was possible to estimate the concentration of virus particles in material by placing drops of various dilutions upon the chorioallantoic membrane and thus to study the growth or the inhibition of growth of the viruses with ease. It was by means of the fertile hen's egg that the advances in virology between 1936 and 1949 were made. Not only were a number of viruses cultivated artificially for the first time, but serological surveys of populations from all over the world enabled the epidemiology of influenza to be worked out in detail. The existence of the three types of virus, A, B and C, was established and the antigenic variation of the A strains which is responsible for their ability to cause successive pandemics was appreciated. The introduction of the fertile hen's egg as a culture medium in virology was comparable in its effects to the introduction of solid media in bacteriology by Robert Koch in the previous century, with this difference – that there remained a number of viruses that would not grow in embryo chick cells.

The development of the electron microscope by Marton in 1934 opened a new era in the study of the morphology of viruses. Even the early instruments increased the resolving power available to microscopists by a factor of ten giving magnifications of up to twenty thousand, and with the best modern machines magnifications of over two hundred thousand are possible. Electron microscope photographs of bacterial viruses were produced within a year or two of the invention of the instrument. By the early nineteen-forties pictures of the elementary bodies of animal viruses were published, such as those of Green of the elementary bodies of *Vaccinia* virus at a magnification of twenty-eight thousand. The later development of heavy-metal shadowing and of ultra-thin sections led to an understanding of the intimate structure of virions. In 1956 Crick and Watson proposed on theoretical grounds that virus particles must be made up of a nucleic acid core and a surrounding shell composed of a number of identical protein subunits. In 1959 such protein subunits were visualized in electron microscope photographs published by Horne and Nagington. The study of the fine structure of viruses continues, new types of virus being looked at each year. More recently the fire structure of the rabies virus has been elucidated. Rather surprisingly it has a bullet-shaped form very similar to that of the bushy stunt virus of tomatoes.

The further development of tissue culture methods of growing viruses was the work of Enders and his colleagues. With the antibiotics penicillin and streptomycin available for the control of bacterial contamination they were able to make rapid progress and by 1949 had shown that poliomyelitis virus could be cultivated in non-neural tissues such as minced monkey kidney. In 1952 Gay and his colleagues established the famous continuous cell line of HeLa cells, derived from a carcinoma of the cervix uterus that occurred in a lady by the name of Helen Lane. In 1953 Scherer and his associates succeeded in growing poliomyelitis virus in a culture of HeLa cells. Thus there were now available both discontinuous and continuous cell lines. The techniques available for the recognition and enumeration of virus growth remained, however, tedious and expensive. The production of vaccines by cultivating attenuated strains of virus in tissue cultures was possible, but the detection of virus in clinical specimens was still not a practical proposition in a diagnostic laboratory.

The publication by Younger in 1954 of a technique for growing trypsinized cells in monolayers upon glass made it possible to recognize viral infection of cells by the detection of the cytopathic effect (CPE). As CPE can be easily observed with a low power microscope ($\times 100$) this has made diagnostic examination of specimens from patients a routine procedure. The identification of unknown viruses by means of the inhibition of the CPE by specific antisera and the examination of patients' sera for the presence of antibodies to a specific virus are also easily carried out using monolayer cultures.

Since 1954 an immense amount of work has been done on the epidemiology of viruses, and live vaccines have been developed against not only poliomyelitis but also measles and rubella. It has proved possible by making use of the intelligence provided by the network of World Health Organization national and international virus laboratories to prepare new influenza vaccines that protect against the mutant strains of influenza A which arise from time to time and cause pandemics. In the case of the Hong Kong strain which was first isolated in 1969 it took only three months from the first isolation of the new strain to the earliest production batch of the vaccine.

The search for antiviral chemotherapeutic agents is at present one of the most active fields of research. There are two approaches to the problem. The first is the testing of a variety of chemical compounds in the hope of finding one that will interfere with some step in replication that is unique to the virus. This approach has led to the discovery of three promising agents. N-methylistatin-β-thiosemicarbazone (Marboran) has been shown to be effective in suppressing the growth of smallpox virus in tissue culture and the results of a field trial in South

India appear to prove its value in the prophylaxis of smallpox; it is however of no value in the treatment of clinical cases (Baurer *et al.* 1963). 5-Iodo-2'-deoxyuridine (IDU) has been shown to be a selective inhibitor of DNA viruses, and is of undoubted value in the treatment of *Herpes simplex* infections involving the cornea in man. There is some evidence that in spite of its toxicity it can be lifesaving when given parenterally in cases of herpetic encephalitis. 1-Adamantanamine has been shown experimentally to inhibit the growth of strains of influenza A virus, though not other myxoviruses, and there is some evidence that it confers protection upon animals and men infected with influenza A virus. Like Marboran, adamantanamine is of little or no use in the therapy of clinical cases. This is not surprising when one remembers that a clinical diagnosis of smallpox or influenza is not possible until the body of the patient is saturated with virus and the disease, in a sense, is more than half over. If we were unable to diagnose lobar pneumonia until the crisis we should be unaware of the therapeutic effects of the sulphonamides or penicillin in this condition.

The other approach to viral chemotherapy has stemmed from the discovery in 1962 by Isaacs of the naturally occurring inhibitor of viral replication, interferon. This substance, which is produced by virus-infected cells, is host-specific but active against all viruses. It appears to be a low molecular weight protein. Experimentally it has been extracted from cells and shown to have both protective and curative effects in tissue culture systems and animals. At present work is going on to devise an economically feasible method for the preparation of interferon on a large scale. Since 1968 a fresh avenue has been opened up by the discovery of compounds which stimulate the synthesis and/or the release of interferon from cells. These compounds fall into two classes: the synthetic double-stranded RNA analogues and antibiotics such as statlon.

REFERENCES

1892 Iwanoski, D. *Bull. Acad. imp. Sci. St Petersberg.* Quoted in: 'Uber die Mosaikkrankheit der Tabakspflanze'. *Z.PflZücht.*, **13**, 1–41.

1898 Copeman, S. M. *Vaccination – Its Natural History and Pathology.* London: Macmillan & Co. Ltd.

1898 Loeffler, F. & Frosch, P. Berichte der Kommission zur Erforschung der Maulund Klauenseuche bei dem Institut für Infektionskrankheiten in Berlin. *Zentralbl.Bakt.*, **23**, 371–91.

1900 Carroll, J. The treatment of yellow fever. *J. Am. med. Ass.* (1902), **39**, 117–24.

1903 Iwanowski, D. Uber die Mosaikkrankheit der Tabakspflanze. *Z. PflZücht.*, **13**, 1–41.

1911 Rous, P. Transmission of a malignant new growth by means of a cell free filtrate. *J. Am. med. Ass.*, **56**, 198.

1915 Noguchi, H. Pure cultivation *in vivo* of vaccine virus free from bacteria. *J. exp. Med.*, **21**, 539–70.

1915 Twort, F. W. An investigation on the nature of ultra-microscopic viruses. *Lancet*, **ii**, 1241–3.

1925 Parker, F. & Nye, R. N. The cultivation of vaccine virus. *Am. J. Path.*, **1**, 325–35.

1925 Parker, F. & Nye, R. N. Cultivation of herpes virus. *Am. J. Path.*, **1**, 337–40.

1925 Barnard, J. E. The microscopical examination of filterable viruses associated with malignant new growths. *Lancet*, **ii**, 117–23.

1927 Carrell, A. & Rivers, T. M. La fabrication du vaccin *in vitro*. *C. r. Séanc. Soc. Biol.*, **96**, 848–50.

1928 Maitland, H. B. & Maitland, M. C. Cultivation of vaccinia virus without tissue culture. *Lancet*, **ii**, 596–7.

1931 Goodpasture, E. W., Woodruff, A. M. & Budding, G. J. The cultivation of vaccine and other viruses in the chorioallantoic membrane of chick embryos. *Science*, **74**, 371–2.

1934 Marton, L. La microscopie electronique des objects biologiques. *Bull. Acad. Belg., Cl. Sci.*, **20**, 439–66.

1935 Stanley, W. M. Isolation of a crystalline protein possessing the properties of tobacco-mosaic virus. *Science*, **81**, 644–5.

1941 Hirst, G. K. The agglutination of red cells by allantoic fluid of chick embryos infected with influenza virus. *Science*, **94**, 22–3.

1942 Hirst, G. K. The quantitative determination of influenza virus and antibiotics by means of red cell agglutination. *J. exp. Med.*, **75**, 49–64.

1949 Enders, J. F., Weller, T. H. & Robbins, F. C. Cultivation of the Lansing strain of poliomyelitis virus in cultures of various human embryonic tissues. *Science*, **109**, 85–7.

1952 Gay, G. O., Coffman, W. D. & Kubicek, M. T. Tissue culture studies of the proliferative capacity of cervical carcinoma and normal epithelium. *Cancer Res.*, **12**, 264–5.

1954 Younger, J. S. Monolayer tissue cultures. I. Preparation and standardization of suspensions of trypsin-dispersed monkey kidney cells. *Proc. Soc. exp. Biol. Med.*, **85**, 202–5.

1956 Crick, F. H. C. & Watson, J. D. Structure of small viruses. *Nature, Lond.*, **177**, 473–5.

1959 Horne, R. W. & Nagington, J. Electron microscope studies of the development and structure of poliomyelitis virus. *J. molec. Biol.*, **1**, 333–8.

1962 Isaacs, A. Production and action of interferon. *Cold Spring Harb. Symp. quant. Biol.*, **27**, 343–9.

1963 Paterson, A., Fox, A. D., Davies, G. *et al.* Controlled studies of IDU in the treatment of hepatic keratitis. *Trans. ophthal. Soc. UK*, **83**, 583–91.

1963 Baurer, D. J., St Vincent, L., Kempe, C. H. & Downie, A. W. Prophylactic treatment of smallpox contacts with *N*-methylisatin-β-thiosemicarbazone (Compound 33T57 Marboran). *Lancet*, **ii**, 494–6.

1965 Tyrell, D. A. J., Bynoe, M. L. & Hoorn, B. Studies on the antiviral activity of 1-adamantanamine. *Br. J. exp. Path.*, **46**, 370–5.

1967 Field, A. K., Tytell, A. A., Lampson, G. P. & Hilleman, M. R. Inducers of interferon and host resistance. II. Multistranded synthetic polynucleotide complexes. *Proc. nat. Acad. Sci. USA*, **58**, 1004–10.

1968 Togo, Y., Hornick, R. B. & Dawkins, A. T. Studies on induced influenza in man. I. Double-blind studies designed to assess prophylactic efficacy of amantadine hydrochloride against A_2/Rockville 1/65 strain. *J. Am. med. Ass.*, **203**, 1089–94.

1968 Banks, G. T., Buck, K. W., Chain, E. B., Himmelweit, F., Marks, J. E., Tyler, J. M., Hollings, M., Last, F. T. & Stone, O. M. Viruses in fungi and interferon stimulation. (The active component of statalon is RNA of viral origin.) *Nature, Lond.*, **218**, 542–5.

1968 Park, J. H. & Baron, S. Herpetic keratoconjunctivitis; therapy with synthetic double-stranded RNA. *Science*, **162**, 811–13.

13

Protozoology

Protozoa were amongst the first microbes to be observed by van Leeuwenhoek of Delft in the sixteen-seventies. Indeed it is clear that he thought that all the microscopic life forms that he described were little animals, for he referred to them as animalcules. His first published descriptions were of free-living forms that he had found in the water of a rainwater barrel. His meticulous drawings enable us to identify *Vorticella*, *Volvox* and *Euglena* amongst other species (van Leeuwenhoek, 1674).

A little later, in 1678, Huygens independently described a number of free-living protozoa including the ciliate *Paramecium*.

To Van Leeuwenhoek also must go the credit for the first description of a pathogenic protozoan, for in 1681 he examined his stools during a bout of diarrhoea and observed *Giardia lamblia* (*intestinalis*) in large numbers. He subsequently discovered similar organisms in the gut of rodents and of frogs.

All over Europe 'ingenious persons' as the seventeenth-century biographer John Aubrey called them, were turning the newly developed microscopes upon a variety of fluids; pond water, beer, blood, semen and excreta. It is not surprising that the end of the seventeenth century should have seen further discoveries of protozoal types. In 1691 the Italian Buonanni, in a monograph published in Rome, gave the first description of *Colpoda*. In 1696 Harris in a paper entitled 'Some microscopical observations of vast numbers of animalcula seen in water' published in the *Philosophical Transactions of the Royal Society*, gave what appears to be an independent description of *Euglena*. It seems strange that he was unaware of van Leeuwenhoek's work, published in the same journal some twenty years before, but keeping up with the literature was perhaps a problem even as early as the seventeenth century!

The first treatise on protozoology was published in 1718 by Louis Joblot, a typical 'virtuoso' of the late seventeenth/early eighteenth century. He was Professor of Mathematics, Geometry and Perspective at the Royal Academy of Painting and Sculpture in Paris and in addition an accomplished microscopist. The work in question is entitled *Descriptions et usages de plusieurs nouveaux microscopes...; avec de nouvelles observations faites sur...animaux de divers espèces, qui naissent dans des liqueurs préparées et dans celles qui ne le sont point,* and consists of two parts, the first dealing with the construction and use of microscopes and the second, illustrated by twelve plates, with the animalcules that he had observed. The first generic name that we retain was '*Paramecium*', coined by Hill in 1752. The term infusoria, to cover the whole group of animalcules was introduced by Wrisberg in 1764. The great Linnaeus, in the later editions of his *Systema Naturae*, placed the infusoria in the class 'Vermes' and made an equivocal contribution to protozoology by the definition of the genus *Chaos*, the type species of which, *C. proteus*, has never been identified by any later observers!

In 1773 in a work devoted to the infusoria and the helminths the Danish biologist Muller first described a species of *Trichomonas* which he had observed in the saliva and scrapings from the teeth of human beings. The same author in his monograph *Animalcula Infusoria Marina*, published posthumously in 1786, produced the first systematic account of the infusoria, describing and naming no less than 379 species.

The term Protozoa was first introduced by the German Goldfuss in a work on the classification of animals, in 1817. It did not immediately replace the older term infusoria for we find that still in use by Ehrenberg in 1838, but by the mid-century it had become standard scientific usage.

Protozoa living in the gut of insects were first noted by Dufour in 1828. *Trichomonas vaginalis*, a flagellate responsible for vaginal discharge in women, was first described by Alfred Donné working at the Hôpital Charité in Paris in 1836.

Thus by 1840 something was known of the free-living Protozoa of fresh and sea water, and symbiotic and parasitic Protozoa colonizing the gut or genital tract of man and animals. It was not, however, until 1841 that the group of Protozoa which live in the blood were discovered. The first report was of the presence of flagellates in the blood of certain salmon by Valentin, a German worker. In the next year, 1842, came a report from Gluge of a peculiar entozoon seen in the blood of frogs. In 1843 David Gruby, a Hungarian physician working in Paris and amongst other things the discoverer of the dermatophytic fungi causing ringworm, proposed the generic name *Trypanosoma* for the micro-organisms described by Gluge. It was, however, to be another fifty years before the species pathogenic for man and domestic animals were discovered.

Throughout the nineteenth century there were a small number of medical microscopists who made a habit of looking at the dejecta, blood, or other material from their patients. It is to these men that we owe the recognition and description of many of the buccal and intestinal protozoa of man. In 1849 Gros described *Entamoeba gingivalis* in material from the mouth. In 1854 Davine examining the stools of patients suffering from cholera observed and described the flagellates *Trichomonas hominum* and *Chilomastix*. In 1857 *Balantidium coli* was first described by Malmsten in the stools of some of his patients. It is important to remember that these observations were all made by direct examination of unstained preparations, neither staining techniques nor artificial culture being available at the time. The remarkable accuracy of the verbal descriptions and drawings of these pioneers is most impressive.

As a part of the European colonial expansion of the last third of the nineteenth century came the development of properly organized medical services in the tropical dependencies of the European powers. Ever since the first 'factories' of the merchant adventurers had been established, a few physicians and ship's surgeons had practised in tropical areas and had made valuable clinical observations upon the diseases peculiar to these zones, but with the establishment of organized colonial medical services (the foremost amongst them being the Indian Medical Service), a degree of specialization became possible, and with the importation of laboratory equipment from Europe an incredibly rich field was opened up for medical parasitologists. Amongst the first fruits of these changes was the discovery in 1870 of *Entamoeba coli* by Lewis of the Indian Medical Service. He found the micro-organism first in the stools of patients suffering from cholera and subsequently in the stools of normal persons. Five years later in 1875 Lösch, working in Russia, first described *Entamoeba histolytica*, the causative organism of amoebic dysentery.

The economic importance of protozoal infections of animals was demonstrated by the work of Louis Pasteur, who investigated the disease, pébrine, which by 1869 had all but ruined the silk-producing industry in the south of France. His report published in 1870 proved that the disease affecting the silkworms was due to infection by a protozoan parasite *Nosema bombycis* and advanced proposals for the control of the infection which proved effective.

Behind the clinical microscopists of Europe or the tropical colonies and those like Pasteur whose aim was the control of industrial or agricultural problems, stood the great German systematic parasitologists collating their observations and defining the higher groups into which the various protozoa described fell. Studies such as that of Eimer on coccidial infections in various animal species and Leuckart's *Die Parasiten des Menschen* (Leipzig, 1879) and Butschli's text *The Protozoa* are typical of this kind of work.

Leuckart's most notable contribution was the definition of the Sporozoa and within it of the Haemosporidia, thus laying down a framework into which the later discovery of the plasmodia of malaria could be easily fitted.

The control of malaria is perhaps the greatest contribution to human welfare that the science of parasitology has made. The story of the observation of the parasites, the description of the various stages in the cycle in man, the discovery of the role of mosquitoes in the transmission of the disease, and thus the development of rational control measures, is one of the most satisfactory in the history of medicine. Major contributions were made by French, Italian and British medical scientists and the work spanned the three continents of, Africa, Asia and Europe. The first observations of parasites in the blood of patients suffering from malaria were made by the French army doctor, Alphonse Lavaran, while he was working in the town of Constantine in Algeria. His findings were published in the *Bulletin de l'Académie de Médecine, Paris* in 1880. Lavaran observed only the merozoites. Over the next nine years the Italian histologist and pathologist Golgi worked out the details of the cycle of development of the parasite in the blood-stream of man and was able to relate the peaks of fever to the occurrence of schizogony and thus recognize that there were at least three species of plasmodium affecting man. Ronald Ross of the Indian Medical Service was the first person to show that malaria parasites underwent another cycle of development in mosquitoes and that the disease was transmitted by the bite of the female anopheline. His first observations in 1898 were made upon avian malaria, but the transmission by the anopheline mosquito was soon confirmed for the human type of infection and this led to the adoption of mosquito eradication, at first by oiling the water where the larvae developed and in more recent times by the use of organic insecticides such as DDT or Gammexane. The immediate dramatic effect of mosquito eradication upon the incidence of malaria ensured that it was rapidly adopted throughout the world wherever malaria was endemic. The long-term results of malaria control and the consequent reduction of infant mortality have yet to be seen, but already the population explosion in many tropical countries to which malaria control has contributed so much, has given rise to a whole new set of problems. The new problems of human population control are soluble in the sense that we possess the technical methods to deal with them, but the social and emotional aspects have so far inhibited any major progress.

Malaria was not the first protozoal disease to be shown to be spread by biting insects. In 1893 Theobald Smith and Kilbourne investigating the causation of Texas cattle fever proved that this disease was caused by a haemoprotozoan, *Babsia*, and that infection was transmitted from one animal to another by the bites of ticks, thus enabling effective control methods to be developed.

In 1891 an important technical advance was made when Romanowsky of St Petersburg introduced his staining method for demonstrating malaria parasites in blood smears. The difficulties under which the pioneers such as Lavaran and Golgi must have worked without the aid of this staining technique were formidable. Ramanowsky's stain made it possible for any physician who possessed a microscope to make a definitive diagnosis of malaria in cases where the disease was suspected. As there was already an extremely effective chemotherapeutic agent available in the form of quinine this improvement in the diagnostic methods available made an important contribution to the well-being of patients.

In 1895 David Bruce, the discoverer of the causative organism of Malta fever, was posted to East Africa, where he became interested in nagana, a widespread condition affecting domestic animals such as bovines and horses and preventing the use of the otherwise suitable grasslands of Zululand from being used for cattle ranching by European settlers. Bruce published two reports upon nagana, one in 1895 and the other in 1897. He showed that the blood of infected animals contained a species of trypanosome, and proved that the infection was transmitted by the bite of the tsetse fly. Control of the insect vector enabled small areas of land to be freed from infection and used for cattle farming, but it soon became apparent that the wild game of the forests and grasslands of Africa were an enormous reservoir of infection, and that eradication of the disease was not possible with the resources available. Indeed even today, with specific chemoprophylaxis and modern insecticides, the condition persists in both East and West Africa and prevents cattle raising being an economic proposition within the tsetse belt.

The human population of the tsetse belt of Africa is subject to the disease called sleeping sickness, a progressive and, if untreated, uniformly fatal neurological disorder characterized by increasing apathy and finally coma and death. In 1902 J. E. Dutton reported finding trypanosomes in the peripheral blood of cases of sleeping sickness. It soon became clear that these organisms were the cause of the disease and that there were two stages; an early one in which the trypanosomes were confined to the bloodstream and a later one in which they could be found in the cerebrospinal fluid and the substance of the brain. The first stage was shown by Robert Koch in 1906 to be curable by atoxyl, but once the parasites have reached the central nervous system the outlook is almost always hopeless.

The disease kala-azar had been known to Indian and Arab physicians for centuries, as it is clinically identifiable at least in its classical form, but no rational theory of its aetiology had been put forward. Then in 1903 W. B. Leishman of the Royal Army Medical Corps (Fig. 13.1) and C. Donovan independently discovered the causative organism, now named in their honour

Fig. 13.1. Sir W. B. Leishman (1865–1926).
Portrait from a photograph. By courtesy
'The Wellcome Trustees'.

Leishmania donovani. Both these authors were working in India and both published their discoveries in the *British Medical Journal.*

Bacteriology and virology both furnish examples of how the introduction of techniques for the isolation and culture in artificial media of pure strains of the organisms were rapidly followed by a dramatic flowering of the subject, giving rise to notable advances in public health and clinical medicine. The first persons to achieve the artificial culture of Protozoa were the Americans W. E. Musgrave and M. T. Clegg working in the Philippines. In 1904 they published an article on 'Amebas: their cultivation and aetiologic significance'. The next year Novy and MacNeal reported the cultivation of avian trypanosomes in sterile blood agar, and since then there have been numerous reports of the in-vitro cultivation of amoebae, trypanosomes and

Leishmania. It is, however, peculiar that no such dramatic advances as occurred in bacteriology and virology have resulted from these technical feats. Microscopic examination of material from the infected animal or man remains the sheet anchor of parasitology whether research or diagnostic.

The year 1919 saw the publication of an account of the first human cases of coccidiosis. The first cases were recognized in Manila, but in later years it became clear that the infection has a wide geographical distribution.

The discovery of the exo-erythrocytic stage in the life cycle of the malaria parasites in man came as a result not of the observations of parasitologists or experts in tropical medicine, but those of psychiatrists. The introduction of malaria treatment for general paralysis of the insane (GPI) by Wagner-Jauregg in 1917 gave those psychiatrists who were using it an unparalleled experience of artificially induced malaria. The precise time when each patient was inoculated with infected blood or bitten by infected mosquitoes was known, as were the precise date and time of the first attack of fever. As more and more of these clinical records were examined it became clear that the accepted account of the behaviour of the plasmodia on first entering the blood-stream would not account for the observed facts. In 1902 the parasitologist Schaudinn, famous as one of the discoverers of the causative organism of syphilis, *Treponema pallidum*, had reported seeing sporozoites which had been injected into the bloodstream entering into erythrocytes and immediately beginning the cycle of intracorpuscular development. This account was universally believed until 1924 when Yorke and Macfie, who were using malarial therapy on GPI patients, reported that if quinine were given immediately before the inoculation of infected blood no infection resulted, but that a single dose given a few hours after the injection of the blood did not prevent infection occurring. This obviously did not fit with Schaudinn's account, but the findings might have been peculiar to infections produced by the artificial method of injecting infected blood. However in 1934 two American psychiatrists, M. F. Boyd and W. K. Stratman-Thomas, reported that the peripheral blood of a patient who had been bitten by fifteen mosquitoes infected with *Plasmodium vivax* as a part of his treatment, did not become infectious on injection into other patients until the ninth day after the bites, and that no parasites could be found in his peripheral blood until the eleventh day. This observation upon a case of infection produced by natural means made it clear that the sporozoites did not, as had been believed, enter directly into the erythrocytes, but were rapidly cleared from the blood-stream and underwent a stage of their development hidden somewhere in the internal organs. Boyd and Stratman-Thomas also showed that the anopheline mosquito was only able to infect man by its bite between about the tenth and the fortieth day after it had fed upon an infected human.

In 1937 A. J. Warren and L. T. Coggleshall working on avian malaria in the USA gave the first clue as to to where the sporozoites went after they had been cleared from the peripheral blood. They reported that while the peripheral blood of the birds was not infectious for the first seventy-two hours after injection of sporozoites, emulsions of liver, spleen and bone-marrow were. It now looked as though the plasmodia were quickly taken up by the reticuloendothelial system and under-went some sort of maturation in those cells before returning to the bloodstream and invading erythrocytes. James and Tait coined the term 'exo-erythrocytic schizogony' for this phase of the development of the parasites in a paper in *Nature* in 1937.

While the war of 1939–1945 led to great advances in our knowledge of the distribution of various protozoal infections in the various tropical theatres of war, and to the development of much more effective drugs for the chemoprophylaxis of malaria, as well as the use of residual spraying with organic insecticides for mosquito control, it inevitably held up the progress of fundamental research. It was not until 1948 that a full knowledge of the exo-erythrocytic cycle of the plasmodia in man was obtained. In that year Shortt and P. C. C. Garnham published their classic paper recording their observations of the exo-erythrocytic forms of the plasmodia in the livers, first of monkeys infected with *Plasmodium cynomologi* and then of men infected with *P. vivax*. They showed that each sporozoite remained in the reticuloendothelial cells of the liver for eight to nine days and that during that time it divided to give rise to about a thousand merozoites which were then released into the bloodstream and entered erythrocytes, initiating the well known erythrocytic cycle of development. Exo-erythrocytic forms have been since observed in all species of plasmodia infecting man, and this phase of development is thought to be universally present in all plasmodia.

Protozoology is one of the fields of medicine where new diseases are still being discovered. In 1939 the first human cases of toxoplasmosis were recorded by Wolf, Cowen and Paige. The parasite, *Toxoplasma gondii*, was first described by the French workers Nicolle and Manceaux in 1908. They had found it in a small rodent, *Ctenodactylus gondi*, indigenous to North Africa. It attracted little attention until human cases were recognized, but since then it has been shown to have a worldwide distribution, and to affect a number of mammals and birds as well as man. The epidemiology of the condition has been worked out using serological tests such as the dye test of Sabin and Feldman, and a toxoplasmin skin test detecting delayed hypersensitivity to the proteins of the parasite.

More recently it has been realized that coprozoic amoebae of the genera *Hartmanella* and *Naegleria* can cause a fatal form of meningo-encephalitis in man. The first species of *Naegleria* was recognized in 1889 by Schardinger, and the first species of *Hartmanella* by Dangeard in 1900. The genera were established by Alexieff in 1912. For many years these free-living amoebae that were sometimes detected in the stools of men and animals were thought to be harmless saprophytes. Their pathogenic potentialities came to be recognized in a rather curious way. In 1930 Castellani reported on certain free-living amoebae that he had found contaminating pure cultures of bacteria. Later in the same year Douglas, who studied and described these organisms, named them *Hartmanella castellani*. In 1956 Jahnes, Fullmer and Li were troubled with amoebic contamination of the monkey kidney tissue cultures that they were using in virological research. The offending organisms proved to be members of the genus *Hartmanella*. Soon other virologists reported similar troubles with contaminating amoebae which caused a cytopathic effect in the tissue cultures. These observations led the American worker Culbertson (Fig. 13.2) to initiate a series of collaborative studies on the experimental pathology of *Hartmanella* infections in mice and monkeys. They found that animals injected with fluid from infected tissue cultures became ill and died: at post-mortem examination they were found to be suffering from

Fig. 13.2. C. D. Culbertson (1906—). From
Am. J. clin. Path. (1961), **35**, facing p. 195.
© Williams & Wilkins Co., Baltimore.

encephalomyelitis, and hartmanellae were demonstrated in the
lesions in the brains and spinal cords of the affected animals. As a
result of a series of such experiments Culbertson and his
colleagues predicted in 1961 that naturally occurring infections
with such coprozooic amoebae would be found if looked for.

In fact the first report of natural infections came from
Australia in 1965 by Fowler and Carter of the Adelaide
Children's Hospital, and their report proved the speculations
founded upon the experimental infections to be well founded.
The free-living amoebae were implicated as the cause of an acute
fatal meningo-encephalitis in four children. Since then cases of
amoebic meningo-encephalitis have been reported from the
United States, Czechoslovakia, New Zealand and Great Britain.
As coprozooic amoebae of the genera *Hartmanella* and *Naegleria*,

both of which can cause these fatal infections, are found in soil, water and air samples from all over the world, it is likely that the naturally occurring infection will be shown to have a global distribution. The story of the discovery of this previously unknown infection is of particular interest as it is, so far as I know, the only case where the recognition of the clinical cases was due to the earlier studies of experimental infections in animals. It is good to record that it has recently been shown that the amoebae are sensitive to amphotericin B, so that there is the possibility of cure if the diagnosis is made early enough.

REFERENCES

1718 Joblot, L. *Descriptions et usages de plusieurs nouveaux microscopes tant simples que composez; avec de nouvelles observations faites sur une multitude innombrable d'insectes, et d'autres animaux de diverses espèces, qui naissent dans des liqueurs préparées et dans celles qui ne le sont point.* Paris: J. Collombat.

1758 Linnaeus, C. *Systema Naturae*, 10th edition.

1786 Muller, O. F. *Animalcula infusoria et marina systematice descripsit et ad vivum delineari curavit O.F.M. opus cura O. Fabricii, 4°, Hauniae.*

1836 Donné, A. Animalcules observés dans les matières purulentes et le produit des sécrétions des organes génitaux de l'homme et de la femme. *C. r. Séanc. Acad. Sci.*, **3**, 865–6.

1838 Ehrenberg, C. G. *Die infusionstierchen als vollkommene Organismen.* Leipzig.

1841 Valentin, G. C. Ueber ein Entozoon im Blute von *Salmo fario. Arch. Anat. Physiol. wiss. Med.*, 435–6.

1843 Gruby, D. Recherches et observations sur une nouvelle espèce d'hématozoaire, *Trypanosoma sanguinis. C. r. Séanc. Acad. Sci.*, **17**, 1134–6.

1849 Gros, G. Fragments d'helminthologie et de physiologie microscopique. *Bull. Soc. imp. Nat. Moscow*, **22**, 549–73.

1857 Malmsten, P. H. Infusonier såsom intestinaldjiur los menniskan. *Hygiea, Stockh.*, **19**, 491–501.

1863 Leuckart, K. G. F. R. Die menschlichen Parasiten und die von ihnen herrübrenden Krankheiten. Leipzig: C. F. Winter. (2 vols.)

1870 Eimer, T. Uber die ei- oder kugelförmigen sogenannten Psorospermien der Wirbelthiere. Ein Beitrag zur Entwickelungsgeschichte der Gregarien und zur Kenntniss dieser Parasiten als Krankheitsursache. Würzburg: A. Stuber.

1870 Pasteur, L. *Etudes sur la maladie des vers à soie.* Paris: Gauthier-Villars. (2 vols.)

1871 Lewis, T. R. On a haematazoon inhabiting human blood. Its relation to chyluria and other diseases. *Ann. Rep. sanit. Comm. India*, **8** (Appendix E), 241–60.

1875 Lösch, F. Massenhafts Entwickelung von Amöben im Dickdarm. *Virchows. Arch. path. Anat. Physiol.*, **65**, 196–211.

1880–1889 Butschli, O. Section on protozoa in H. G. Bronn's *Klassen und Ordnungen des Tier-reichs*. Leipzig: C. F. Winter. (3 vols.)

1881 Laveran, C. L. A. Un nouveau parasite trouvé dans la sang de plusieurs malades atteints de fièvre palustre. *Bull. Soc. méd. Hôp. Paris (Mém.)*, 2ᵉ sér., **17**, 158–64.

1895 Bruce, D. Preliminary report on the tsetse fly disease or nagana in Zululand. Durban: Bennett & Davis.

1897 Ross, R. On some peculiar pigmented cells found in two mosquitoes fed on malarial blood. *Br. med. J.*, **ii**, 1786–8.

1898 Ross, R. Report on a preliminary investigation into malaria in the Sigur Ghat, Ootacamund. *Indian Lancet, Calcutta*, **12**, 269–71.

1902 Dutton, J. E. Preliminary note upon a trypanosome occurring in the blood of man. *Thompson Yates Lab. Rep.*, **4**, 455–68.

1903 Leishman, W. B. On the possibility of the occurrence of trypanosomiasis in India. *Br. med. J.*, **i**, 1252–4.

1903 Donovan, C. On the possibility of the occurrence of trypanosomiasis in India. *Br. med. J.*, **ii**, 79.

1904 Musgrave, W. E. & Clegg, M. T. Part 1: Amebas: their cultivation and etiologic significance. Part 2: Treatment of intestinal amebas (amebic dysentery) in the tropics. *Dept Interior Govt. Biol. Lab. Bull. Manila*, **18**.

1905 Novy, F. G. & MacNeal, W. J. On the trypanosomes of birds. *J. infect. Dis.*, **2**, 256–308.

1908 Nicolle, C. J. H. & Manceaux, L. H. Sur une infection à corps de Leishman (ou organismes voisins) du gondi. *C. r. Séanc. Acad. Sci.*, **147**, 763–6.

1918 Haughwout, F. G. The Protozoa of Manila, and the vicinity. *Philippine J. Sci.*, Ser. D, **13**, 175–214.

1922 Wagner-Jauregg, J. Die Malariabehandlung der progressiven Paralyse. *J. nerv. ment. Dis., NY*, **55**, 369–75.

1924 Yorke, W. & Macfie, J. W. S. Certain observations on malaria made during treatment of general paralysis. *Trans. R. Soc. trop. Med. Hyg.*, **18**, 13–44.

1930 Castellani, A. An amoeba found in culture of a yeast: preliminary note. *J. trop. Med. Hyg.*, **33**, 160.

1930 Douglas, M. Notes on the classification of the amoeba found by Castellani in cultures of a yeast-like fungus. *J. trop. Med. Hyg.*, **33**, 258–9.

1934 Boyd, M. F. & Stratman-Thomas, W. K. on the duration of infectiousness in anophelines harboring *Plasmodium vivax. Am. J. Hyg.*, **19**, 539–40.

1937 Warren, A. J. & Coggleshall, L. T. Infectivity of blood and organs in canaries after inoculation with sporozoites. *Am. J. Hyg.*, **26**, 1–10.

1937 James, S. P. & Tait, P. Newer knowledge of the life cycle of malaria parasites. *Nature, Lond.*, **139**, 545.

1939 Wolf, A., Cowen, D. & Paige, D. H. A new case of granulomatous encephalitis due to a protozoon. *Am. J. Hyg.*, **15**, 657–94.

1948 Shortt, H. E. & Garnham, P. C. C. Exoerythrocytic parasites of *Plasmodium cynomolgi. Trans. R. Soc. trop. Med. Hyg.*, **41**, 485–795.

1957 Jahnes, W. G., Fullmer, H. M. & Li, C. P. Free living amoebae as contaminants in monkey kidney tissue culture (23515). *Proc. Soc. exp. Biol. Med.*, **96**, 484–8.

1958 Culbertson, C. G., Smith, J. W. & Minner, J. R. *Acanthamoeba*: observations on animal pathology. *Science*, **127**, 1506.

1959 Culbertson, C. G., Smith, J. W., Cohen, H. K. & Minner, J. R. Experimental infection of mice and monkeys by *Acanthamoeba. Am. J. Path.*, **35**, 185–98.

1961 Culbertson, C. G., Overton, W. M. & Reveal, M. A. Pathogenic *Acanthamoeba (Hartmanella). Am. J. clin. Path.*, **35**, 195–202.

1965 Culbertson, C. G., Holmes, D. H. & Overton, W. M. *Hartmanella castellani (Acanthamoeba* sp.) *Am. J. clin. Path.*, **43**, 361–4.

1965 Fowler, M. & Carter, R. F. Acute pyogenic meningitis probably due to *Acanthamoeba* sp.: a preliminary report. *Br. med. J.*, **ii**, 740–2.

1966 Culbertson, C. G., Ensminger, P. W. & Overton, W. M. *Hartmanella (Acanthamoeba). Am. J. clin. Path.*, **46**, 305–14.

1968 Culbertson, C. G., Ensminger, P. W. & Overton, W. M. Pathogenic *Naegleria* sp. – study of a strain isolated from human cerebrospinal fluid. *J. Protozool.*, **15**, 355–63.

1968 Červa, L., Novák, K. & Culbertson, C. G. An outbreak of acute, fatal, amebic meningoencephalitis. *Am. J. Epid.*, **88**, 436–44.

1968 Červa, L. & Novák, K. Amoebic meningoencephalitis: sixteen fatalities. *Science*, **160**, 92.

14

A note on mycology

Microbiology overlaps with mycology. Only a part of the subject is properly the concern of microbiologists; the microscopic as opposed to the macroscopic fungi. Mushrooms and toadstools are quite clearly central to mycology but excluded by their size from microbiology. This note will concentrate upon the development of microbiological mycology and medical mycology in particular. It is a bare outline included in order to compare the development of medical mycology with that of bacteriology and virology.

Whilst the larger fungi have been studied since classical times and descriptions of the various species published with the laudable aim of differentiating between the edible and the poisonous types, the microscopic fungi were not recognized until the late seventeenth century. In 1665 in his *Micrographia* Robert Hooke published the first illustrations of microscopic fungi. Antony van Leeuwenhoek in one of his letters to the Royal Society in 1667 gave the first description of the yeast cells in beer wort. In 1679 Malpighi published drawings of moulds recognizable as *Rhizopus, Mucor* and *Penicillium,* in his *Anatome Plantarum.*

Perhaps the most important pioneer in the field was Pietro Antonio Micheli, a Tuscan botanist who was in charge of the public gardens of Florence during the early eighteenth century. In 1729 he published his *Nova Genera Plantarum,* a monumental work. Amongst the 1900 genera enumerated 900 were fungi. It is to him that we owe the naming and definition of *Mucor* and *Aspergillus,* amongst other genera. Even more important he was the first person to recognize and describe fungal spores, which he referred to as seeds. His techniques were far ahead of his time and he was the first person to obtain pure cultures of fungi using solid media. He inoculated spores from different organisms onto the surfaces of truncated pyramids of melon, pear and quince. He even reported on the occurrence of contamination of his cultures, which he rightly regarded as being due to spores of adventitious organisms falling onto the cut surface from the air. His methods of study were not equalled again until the work of Robert Koch in the mid nineteenth century.

The pathogenic potentialities of fungi were not appreciated until the latter half of the eighteenth century. Plant pathogens were the first to be recognized. In 1767 Trogioni-Tozetti advanced the hypothesis that the rust diseases of cereals might be caused by microscopic fungi; he did not however adduce any observational or experimental evidence to support his theory. It was not until 1775 that the first positive evidence implicating fungi in plant diseases was published, when Mathieu Tillet, a French civil servant, proved by experiments that bunt of wheat (caused by a smut fungus) was contagious, and in addition showed how it could be controlled by treatment of the seeds. Final proof of the role of microscopic fungi in rust diseases came in 1807 when I. B. Prevost published a monograph in which he described experimental smut infections and illustrated the germination of smut spores. He also showed how the disease could be prevented by steeping infected seeds in copper sulphate solution. At the time Prevost's work did not attract any attention from the agricultural community, but its importance was recognized forty years later after the Irish potato famine had forced farmers to consider the cause and control of fungal infections of crops.

Fungal infections of animals were recognized even later than those of plants. The first to be recorded was avian aspergillosis, which was reported in the lungs of a diseased flamingo by Richard Owen in 1832. In 1837 came Augustino Bassi's classical work in which he demonstrated that the cause of the 'muscardine' disease of silkworms was due to infection with a microscopic fungus.

Ringworm was the first human disease to be shown to be caused by a fungal infection. In 1839 Schoenlein reported finding fungal elements in lesions, and between 1841 and 1843 the species associated with the common human ringworm infections were described in detail by David Gruby working in

Paris. *Microsporon andouni* was described in 1843 and *Trichophyton tonsurans* in 1844. Gruby, a Hungarian of Jewish descent, was primarily a physiologist. He was the teacher of the great Claude Bernard, but also practised as a physician in Paris where he had amongst his patients Alexandre Dumas the elder, George Sand, Chopin, Liszt, and the German poet Heine. In his old age he lived the life of a recluse and devoted himself to clockmaking and astronomy.

The next condition to be shown to be of fungal origin was thrush. *Candida albicans* was described as the causative organism by the Swede F. T. Berg in 1844.

The recognition of the deep and systemic mycoses of man was delayed for many years. The first to be studied was Madura foot, so called because it was first reported in the state of Madura, India, which is now known to be a mycetoma. This condition, though a deep mycosis affecting muscles, tendons and bone, also ulcerated through the skin and therefore the lesions were suitable for material to be removed and examined microscopically. The clinical entity was well described in 1860 by Van Dyke Carter of the Bombay Medical Service. In 1894 Vincent made further studies of the disease and named the organism that he observed in the lesions and which he regarded as the causal organism *Actinomyces madurae*. Up to this point the condition was thought to be limited to India, but in 1916 Chalmers and Archibald published an account of cases seen in the Sudan. They also re-investigated the aetiology and came to the conclusion that the condition was caused not by an actinomycete but by a variety of fungi; hence they named the condition mycetoma.

In spite of the lead given by the work of Micheli in the eighteenth century, medical mycologists were slow to apply cultural techniques to the isolation and characterization of the organisms that they observed in the lesions of the mycoses and this naturally retarded progress. It was not until 1910 that Sabouraud introduced suitable media into medical mycological practice. He was at first misled by the pleomorphism of the fungi and in his monograph he distinguishes between forty-five different species of dermatophytic fungi. The number of species that were acceptable was soon drastically reduced, but the gain to medical mycology from the introduction of pure culture methods has been enormous.

About the turn of the century the systemic mycoses began to be recognized. First in 1892 coccidiomycosis in man was described in Argentina by Posada, then in 1906 histoplasmosis was recognized by Darling during histological work on kala-azar. Darling believed the organisms he had seen to be protozoa, but in 1912 Da Rocha Lima reported budding by the organism and gave it as his opinion that it was a cryptococcus. It was not until 1934 that the issue was finally settled by Monbreun who isolated *Histoplasma capsulatum* in pure culture, showed that it was a fungus, and produced experimental infections in dogs by inoculating the pure cultures.

Mycology remains the most backward of the branches of medical microbiology, partly because fungal infections are much more common in the tropical and subtropical underdeveloped countries of the world, where facilities for research are severely limited by economic factors. Nonetheless, progress is being made, the distribution of known mycoses is being better defined, and new infections are being recognized.

The growing points of mycological research in the industrialized countries are two. In the first place there is much interest in industrial fermentations for the production of antibiotics and steroids as well as the simpler organic compounds. In the second place there is increasing interest in the mycotoxicoses, disease of animals and potentially of man caused by the ingestion of poisonous substances elaborated by fungi contaminating stored food crops. Ergotism has of course been well known for many years, but it is now very rare, and the renewed interest stems from the discovery of aflatoxin, a substance produced by *Aspergillus flavus* contaminating stored groundnuts, and originally isolated in 1960 during an investigation of an outbreak of turkey X disease in Britain. Apart from causing the turkey disease aflatoxin has been shown to cause carcinoma of the liver in rats that have been given minute quantities. In the last few years an active search for other mycotoxins, which may be the cause of various ill-understood diseases of animals, has been in progress and looks like yielding some interesting results.

FURTHER READING

1938 Bulloch, W. *The History of Bacteriology*. London: Oxford University Press.

1940 Large, E. C. *The Advance of the Fungi*. New York: H. Holt & Co., 480 pp.

1941 Ramsbottom, J. Presidential Address: 'The expanding knowledge of mycology since Linnaeus'. *Proc. Linn. Soc. Lond.*, **151**, part 4, 280–367.

1941 Reed, H. S. *A Short History of the Plant Sciences*, p. 323 (New Series of Plant Science Books 7). Waltham, Mass.: Chronica Botanica Co.

1944 Zakon, S. J. & Benedek, T. David Gruby and the centenary of medical mycology, 1841–1941. *Bull. Hist. Med.*, **16**, 155–68.

1976 Ainsworth, G. C. *Introduction to the History of Mycology*. London: Cambridge University Press.

INDEX OF NAMES

SUBJECT INDEX

196